Table of Contents

Think Python

Allen B. Downey

O'REILLY®

Beijing · Cambridge · Farnham · Köln · Sebastopol · Tokyo

Think Python

by Allen B. Downey

Copyright © 2012 Allen Downey. All rights reserved.

Printed in the United States of America.

Published by O'Reilly Media, Inc., 1005 Gravenstein Highway North, Sebastopol, CA 95472.

O'Reilly books may be purchased for educational, business, or sales promotional use. Online editions are also available for most titles (*http://my.safaribooksonline.com*). For more information, contact our corporate/institutional sales department: 800-998-9938 or *corporate@oreilly.com*.

Editors: Mike Loukides and Meghan Blanchette
Production Editor: Rachel Steely

Proofreader: Stacie Arellano
Cover Designer: Karen Montgomery
Interior Designer: David Futato
Illustrators: Robert Romano and Rebecca Demarest

August 2012:   First Edition

Revision History for the First Edition:

2012-08-03   First release

See *http://oreilly.com/catalog/errata.csp?isbn=9781449330729* for release details.

ISBN: 978-1-449-33072-9

[LSI]

Preface

The Strange History of This Book

In January 1999 I was preparing to teach an introductory programming class in Java. I had taught it three times and I was getting frustrated. The failure rate in the class was too high and, even for students who succeeded, the overall level of achievement was too low.

One of the problems I saw was the books. They were too big, with too much unnecessary detail about Java, and not enough high-level guidance about how to program. And they all suffered from the trap door effect: they would start out easy, proceed gradually, and then somewhere around Chapter 5 the bottom would fall out. The students would get too much new material, too fast, and I would spend the rest of the semester picking up the pieces.

Two weeks before the first day of classes, I decided to write my own book. My goals were:

- Keep it short. It is better for students to read 10 pages than not read 50 pages.
- Be careful with vocabulary. I tried to minimize the jargon and define each term at first use.
- Build gradually. To avoid trap doors, I took the most difficult topics and split them into a series of small steps.
- Focus on programming, not the programming language. I included the minimum useful subset of Java and left out the rest.

I needed a title, so on a whim I chose *How to Think Like a Computer Scientist*.

My first version was rough, but it worked. Students did the reading, and they understood enough that I could spend class time on the hard topics, the interesting topics and (most important) letting the students practice.

I released the book under the GNU Free Documentation License, which allows users to copy, modify, and distribute the book.

What happened next is the cool part. Jeff Elkner, a high school teacher in Virginia, adopted my book and translated it into Python. He sent me a copy of his translation, and I had the unusual experience of learning Python by reading my own book. As Green Tea Press, I published the first Python version in 2001.

In 2003 I started teaching at Olin College and I got to teach Python for the first time. The contrast with Java was striking. Students struggled less, learned more, worked on more interesting projects, and generally had a lot more fun.

Over the last nine years I continued to develop the book, correcting errors, improving some of the examples and adding material, especially exercises.

The result is this book, now with the less grandiose title *Think Python*. Some of the changes are:

- I added a section about debugging at the end of each chapter. These sections present general techniques for finding and avoiding bugs, and warnings about Python pitfalls.

- I added more exercises, ranging from short tests of understanding to a few substantial projects. And I wrote solutions for most of them.

- I added a series of case studies—longer examples with exercises, solutions, and discussion. Some are based on Swampy, a suite of Python programs I wrote for use in my classes. Swampy, code examples, and some solutions are available from *http://thinkpython.com*.

- I expanded the discussion of program development plans and basic design patterns.

- I added appendices about debugging, analysis of algorithms, and UML diagrams with Lumpy.

I hope you enjoy working with this book, and that it helps you learn to program and think, at least a little bit, like a computer scientist.

—Allen B. Downey
Needham, MA

Acknowledgments

Many thanks to Jeff Elkner, who translated my Java book into Python, which got this project started and introduced me to what has turned out to be my favorite language.

Thanks also to Chris Meyers, who contributed several sections to *How to Think Like a Computer Scientist*.

Thanks to the Free Software Foundation for developing the GNU Free Documentation License, which helped make my collaboration with Jeff and Chris possible, and Creative Commons for the license I am using now.

Thanks to the editors at Lulu who worked on *How to Think Like a Computer Scientist*.

Thanks to all the students who worked with earlier versions of this book and all the contributors (listed below) who sent in corrections and suggestions.

Contributor List

More than 100 sharp-eyed and thoughtful readers have sent in suggestions and corrections over the past few years. Their contributions, and enthusiasm for this project, have been a huge help. If you have a suggestion or correction, please send email to feed back@thinkpython.com. If I make a change based on your feedback, I will add you to the contributor list (unless you ask to be omitted).

If you include at least part of the sentence the error appears in, that makes it easy for me to search. Page and section numbers are fine, too, but not quite as easy to work with. Thanks!

- Lloyd Hugh Allen sent in a correction to Section 8.4.
- Yvon Boulianne sent in a correction of a semantic error in Chapter 5.
- Fred Bremmer submitted a correction in Section 2.1.
- Jonah Cohen wrote the Perl scripts to convert the LaTeX source for this book into beautiful HTML.
- Michael Conlon sent in a grammar correction in Chapter 2 and an improvement in style in Chapter 1, and he initiated discussion on the technical aspects of interpreters.
- Benoit Girard sent in a correction to a humorous mistake in Section 5.6.
- Courtney Gleason and Katherine Smith wrote horsebet.py, which was used as a case study in an earlier version of the book. Their program can now be found on the website.
- Lee Harr submitted more corrections than we have room to list here, and indeed he should be listed as one of the principal editors of the text.
- James Kaylin is a student using the text. He has submitted numerous corrections.

- David Kershaw fixed the broken `catTwice` function in Section 3.10.

- Eddie Lam has sent in numerous corrections to Chapters 1, 2, and 3. He also fixed the Makefile so that it creates an index the first time it is run and helped us set up a versioning scheme.

- Man-Yong Lee sent in a correction to the example code in Section 2.4.

- David Mayo pointed out that the word "unconsciously" in Chapter 1 needed to be changed to "subconsciously."

- Chris McAloon sent in several corrections to Sections 3.9 and 3.10.

- Matthew J. Moelter has been a long-time contributor who sent in numerous corrections and suggestions to the book.

- Simon Dicon Montford reported a missing function definition and several typos in Chapter 3. He also found errors in the `increment` function in Chapter 13.

- John Ouzts corrected the definition of "return value" in Chapter 3.

- Kevin Parks sent in valuable comments and suggestions as to how to improve the distribution of the book.

- David Pool sent in a typo in the glossary of Chapter 1, as well as kind words of encouragement.

- Michael Schmitt sent in a correction to the chapter on files and exceptions.

- Robin Shaw pointed out an error in Section 13.1, where the printTime function was used in an example without being defined.

- Paul Sleigh found an error in Chapter 7 and a bug in Jonah Cohen's Perl script that generates HTML from LaTeX.

- Craig T. Snydal is testing the text in a course at Drew University. He has contributed several valuable suggestions and corrections.

- Ian Thomas and his students are using the text in a programming course. They are the first ones to test the chapters in the latter half of the book, and they have made numerous corrections and suggestions.

- Keith Verheyden sent in a correction in Chapter 3.

- Peter Winstanley let us know about a longstanding error in our Latin in Chapter 3.

- Chris Wrobel made corrections to the code in the chapter on file I/O and exceptions.

- Moshe Zadka has made invaluable contributions to this project. In addition to writing the first draft of the chapter on Dictionaries, he provided continual guidance in the early stages of the book.

- Christoph Zwerschke sent several corrections and pedagogic suggestions, and explained the difference between *gleich* and *selbe*.

- James Mayer sent us a whole slew of spelling and typographical errors, including two in the contributor list.

- Hayden McAfee caught a potentially confusing inconsistency between two examples.

- Angel Arnal is part of an international team of translators working on the Spanish version of the text. He has also found several errors in the English version.

- Tauhidul Hoque and Lex Berezhny created the illustrations in Chapter 1 and improved many of the other illustrations.

- Dr. Michele Alzetta caught an error in Chapter 8 and sent some interesting pedagogic comments and suggestions about Fibonacci and Old Maid.

- Andy Mitchell caught a typo in Chapter 1 and a broken example in Chapter 2.

- Kalin Harvey suggested a clarification in Chapter 7 and caught some typos.

- Christopher P. Smith caught several typos and helped us update the book for Python 2.2.

- David Hutchins caught a typo in the Foreword.

- Gregor Lingl is teaching Python at a high school in Vienna, Austria. He is working on a German translation of the book, and he caught a couple of bad errors in Chapter 5.

- Julie Peters caught a typo in the Preface.

- Florin Oprina sent in an improvement in `makeTime`, a correction in `printTime`, and a nice typo.

- D. J. Webre suggested a clarification in Chapter 3.

- Ken found a fistful of errors in Chapters 8, 9 and 11.

- Ivo Wever caught a typo in Chapter 5 and suggested a clarification in Chapter 3.

- Curtis Yanko suggested a clarification in Chapter 2.

- Ben Logan sent in a number of typos and problems with translating the book into HTML.

- Jason Armstrong saw the missing word in Chapter 2.

- Louis Cordier noticed a spot in Chapter 16 where the code didn't match the text.

- Brian Cain suggested several clarifications in Chapters 2 and 3.

- Rob Black sent in a passel of corrections, including some changes for Python 2.2.

- Jean-Philippe Rey at Ecole Centrale Paris sent a number of patches, including some updates for Python 2.2 and other thoughtful improvements.

- Jason Mader at George Washington University made a number of useful suggestions and corrections.

- Jan Gundtofte-Bruun reminded us that "a error" is an error.

- Abel David and Alexis Dinno reminded us that the plural of "matrix" is "matrices", not "matrixes." This error was in the book for years, but two readers with the same initials reported it on the same day. Weird.

- Charles Thayer encouraged us to get rid of the semi-colons we had put at the ends of some statements and to clean up our use of "argument" and "parameter."

- Roger Sperberg pointed out a twisted piece of logic in Chapter 3.

- Sam Bull pointed out a confusing paragraph in Chapter 2.

- Andrew Cheung pointed out two instances of "use before def."

- C. Corey Capel spotted the missing word in the Third Theorem of Debugging and a typo in Chapter 4.

- Alessandra helped clear up some Turtle confusion.

- Wim Champagne found a brain-o in a dictionary example.

- Douglas Wright pointed out a problem with floor division in `arc`.

- Jared Spindor found some jetsam at the end of a sentence.

- Lin Peiheng sent a number of very helpful suggestions.

- Ray Hagtvedt sent in two errors and a not-quite-error.

- Torsten Hübsch pointed out an inconsistency in Swampy.

- Inga Petuhhov corrected an example in Chapter 14.

- Arne Babenhauserheide sent several helpful corrections.

- Mark E. Casida is is good at spotting repeated words.

- Scott Tyler filled in a that was missing. And then sent in a heap of corrections.

- Gordon Shephard sent in several corrections, all in separate emails.

- Andrew Turner spotted an error in Chapter 8.

- Adam Hobart fixed a problem with floor division in `arc`.

- Daryl Hammond and Sarah Zimmerman pointed out that I served up `math.pi` too early. And Zim spotted a typo.

- George Sass found a bug in a Debugging section.

- Brian Bingham suggested Exercise 11-10.

- Leah Engelbert-Fenton pointed out that I used `tuple` as a variable name, contrary to my own advice. And then found a bunch of typos and a "use before def."

- Joe Funke spotted a typo.

- Chao-chao Chen found an inconsistency in the Fibonacci example.

- Jeff Paine knows the difference between space and spam.

- Lubos Pintes sent in a typo.

- Gregg Lind and Abigail Heithoff suggested Exercise 14-4.

- Max Hailperin has sent in a number of corrections and suggestions. Max is one of the authors of the extraordinary *Concrete Abstractions*, which you might want to read when you are done with this book.

- Chotipat Pornavalai found an error in an error message.

- Stanislaw Antol sent a list of very helpful suggestions.

- Eric Pashman sent a number of corrections for Chapters 4–11.

- Miguel Azevedo found some typos.

- Jianhua Liu sent in a long list of corrections.

- Nick King found a missing word.

- Martin Zuther sent a long list of suggestions.

- Adam Zimmerman found an inconsistency in my instance of an "instance" and several other errors.

- Ratnakar Tiwari suggested a footnote explaining degenerate triangles.

- Anurag Goel suggested another solution for is_abecedarian and sent some additional corrections. And he knows how to spell Jane Austen.

- Kelli Kratzer spotted one of the typos.

- Mark Griffiths pointed out a confusing example in Chapter 3.

- Roydan Ongie found an error in my Newton's method.

- Patryk Wolowiec helped me with a problem in the HTML version.

- Mark Chonofsky told me about a new keyword in Python 3.

- Russell Coleman helped me with my geometry.

- Wei Huang spotted several typographical errors.

- Karen Barber spotted the the oldest typo in the book.

- Nam Nguyen found a typo and pointed out that I used the Decorator pattern but didn't mention it by name.

- Stéphane Morin sent in several corrections and suggestions.

- Paul Stoop corrected a typo in uses_only.

- Eric Bronner pointed out a confusion in the discussion of the order of operations.

- Alexandros Gezerlis set a new standard for the number and quality of suggestions he submitted. We are deeply grateful!

- Gray Thomas knows his right from his left.

- Giovanni Escobar Sosa sent a long list of corrections and suggestions.

- Alix Etienne fixed one of the URLs.

- Kuang He found a typo.

- Daniel Neilson corrected an error about the order of operations.

- Will McGinnis pointed out that polyline was defined differently in two places.

- Swarup Sahoo spotted a missing semi-colon.

- Frank Hecker pointed out an exercise that was under-specified, and some broken links.

- Animesh B helped me clean up a confusing example.

- Martin Caspersen found two round-off errors.
- Gregor Ulm sent several corrections and suggestions.

## CHAPTER 1

The Way of the Program

The goal of this book is to teach you to think like a computer scientist. This way of thinking combines some of the best features of mathematics, engineering, and natural science. Like mathematicians, computer scientists use formal languages to denote ideas (specifically computations). Like engineers, they design things, assembling components into systems and evaluating tradeoffs among alternatives. Like scientists, they observe the behavior of complex systems, form hypotheses, and test predictions.

The single most important skill for a computer scientist is **problem solving**. Problem solving means the ability to formulate problems, think creatively about solutions, and express a solution clearly and accurately. As it turns out, the process of learning to program is an excellent opportunity to practice problem-solving skills. That's why this chapter is called, "The way of the program."

On one level, you will be learning to program, a useful skill by itself. On another level, you will use programming as a means to an end. As we go along, that end will become clearer.

The Python Programming Language

The programming language you will learn is Python. Python is an example of a **high-level language**; other high-level languages you might have heard of are C, C++, Perl, and Java.

There are also **low-level languages**, sometimes referred to as "machine languages" or "assembly languages." Loosely speaking, computers can only run programs written in low-level languages. So programs written in a high-level language have to be processed before they can run. This extra processing takes some time, which is a small disadvantage of high-level languages.

The advantages are enormous. First, it is much easier to program in a high-level language. Programs written in a high-level language take less time to write, they are shorter and easier to read, and they are more likely to be correct. Second, high-level languages are **portable**, meaning that they can run on different kinds of computers with few or no modifications. Low-level programs can run on only one kind of computer and have to be rewritten to run on another.

Due to these advantages, almost all programs are written in high-level languages. Low-level languages are used only for a few specialized applications.

Two kinds of programs process high-level languages into low-level languages: **interpreters** and **compilers**. An interpreter reads a high-level program and executes it, meaning that it does what the program says. It processes the program a little at a time, alternately reading lines and performing computations. Figure 1-1 shows the structure of an interpreter.

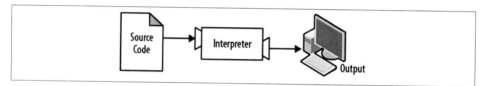

Figure 1-1. An interpreter processes the program a little at a time, alternately reading lines and performing computations.

A compiler reads the program and translates it completely before the program starts running. In this context, the high-level program is called the **source code**, and the translated program is called the **object code** or the **executable**. Once a program is compiled, you can execute it repeatedly without further translation. Figure 1-2 shows the structure of a compiler.

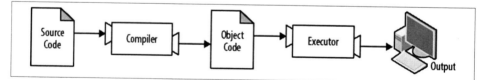

Figure 1-2. A compiler translates source code into object code, which is run by a hardware executor.

Python is considered an interpreted language because Python programs are executed by an interpreter. There are two ways to use the interpreter: **interactive mode** and **script mode**. In interactive mode, you type Python programs and the interpreter displays the result:

```
>>> 1 + 1
2
```

The chevron, >>>, is the **prompt** the interpreter uses to indicate that it is ready. If you type 1 + 1, the interpreter replies 2.

Alternatively, you can store code in a file and use the interpreter to execute the contents of the file, which is called a **script**. By convention, Python scripts have names that end with `.py`.

To execute the script, you have to tell the interpreter the name of the file. If you have a script named `dinsdale.py` and you are working in a UNIX command window, you type `python dinsdale.py`. In other development environments, the details of executing scripts are different. You can find instructions for your environment at the Python website *http://python.org*.

Working in interactive mode is convenient for testing small pieces of code because you can type and execute them immediately. But for anything more than a few lines, you should save your code as a script so you can modify and execute it in the future.

What Is a Program?

A **program** is a sequence of instructions that specifies how to perform a computation. The computation might be something mathematical, such as solving a system of equations or finding the roots of a polynomial, but it can also be a symbolic computation, such as searching and replacing text in a document or (strangely enough) compiling a program.

The details look different in different languages, but a few basic instructions appear in just about every language:

input:
    Get data from the keyboard, a file, or some other device.

output:
    Display data on the screen or send data to a file or other device.

math:
    Perform basic mathematical operations like addition and multiplication.

conditional execution:
    Check for certain conditions and execute the appropriate code.

repetition:
    Perform some action repeatedly, usually with some variation.

Believe it or not, that's pretty much all there is to it. Every program you've ever used, no matter how complicated, is made up of instructions that look pretty much like these. So you can think of programming as the process of breaking a large, complex task into smaller and smaller subtasks until the subtasks are simple enough to be performed with one of these basic instructions.

That may be a little vague, but we will come back to this topic when we talk about **algorithms**.

What Is Debugging?

Programming is error-prone. For whimsical reasons, programming errors are called **bugs** and the process of tracking them down is called **debugging**.

Three kinds of errors can occur in a program: syntax errors, runtime errors, and semantic errors. It is useful to distinguish between them in order to track them down more quickly.

Syntax Errors

Python can only execute a program if the syntax is correct; otherwise, the interpreter displays an error message. **Syntax** refers to the structure of a program and the rules about that structure.For example, parentheses have to come in matching pairs, so (1 + 2) is legal, but 8) is a **syntax error**.

In English readers can tolerate most syntax errors, which is why we can read the poetry of e. e. cummings without spewing error messages. Python is not so forgiving. If there is a single syntax error anywhere in your program, Python will display an error message and quit, and you will not be able to run your program. During the first few weeks of your programming career, you will probably spend a lot of time tracking down syntax errors. As you gain experience, you will make fewer errors and find them faster.

Runtime Errors

The second type of error is a runtime error, so called because the error does not appear until after the program has started running. These errors are also called **exceptions** because they usually indicate that something exceptional (and bad) has happened.

Runtime errors are rare in the simple programs you will see in the first few chapters, so it might be a while before you encounter one.

Semantic Errors

The third type of error is the **semantic error**. If there is a semantic error in your program, it will run successfully in the sense that the computer will not generate any error messages, but it will not do the right thing. It will do something else. Specifically, it will do what you told it to do.

The problem is that the program you wrote is not the program you wanted to write. The meaning of the program (its semantics) is wrong. Identifying semantic errors can be tricky because it requires you to work backward by looking at the output of the program and trying to figure out what it is doing.

Experimental Debugging

One of the most important skills you will acquire is debugging. Although it can be frustrating, debugging is one of the most intellectually rich, challenging, and interesting parts of programming.

In some ways, debugging is like detective work. You are confronted with clues, and you have to infer the processes and events that led to the results you see.

Debugging is also like an experimental science. Once you have an idea about what is going wrong, you modify your program and try again. If your hypothesis was correct, then you can predict the result of the modification, and you take a step closer to a working program. If your hypothesis was wrong, you have to come up with a new one. As Sherlock Holmes pointed out, "When you have eliminated the impossible, whatever remains, however improbable, must be the truth." (A. Conan Doyle, *The Sign of Four*)

For some people, programming and debugging are the same thing. That is, programming is the process of gradually debugging a program until it does what you want. The idea is that you should start with a program that does *something* and make small modifications, debugging them as you go, so that you always have a working program.

For example, Linux is an operating system that contains thousands of lines of code, but it started out as a simple program Linus Torvalds used to explore the Intel 80386 chip. According to Larry Greenfield, "One of Linus's earlier projects was a program that would switch between printing AAAA and BBBB. This later evolved to Linux." (*The Linux Users' Guide* Beta Version 1).

Later chapters will make more suggestions about debugging and other programming practices.

Formal and Natural Languages

Natural languages are the languages people speak, such as English, Spanish, and French. They were not designed by people (although people try to impose some order on them); they evolved naturally.

Formal languages are languages that are designed by people for specific applications. For example, the notation that mathematicians use is a formal language that is particularly good at denoting relationships among numbers and symbols. Chemists use a formal language to represent the chemical structure of molecules. And most importantly:

> **Programming languages are formal languages that have been designed to express computations.**

Formal languages tend to have strict rules about syntax. For example, $3 + 3 = 6$ is a syntactically correct mathematical statement, but $3+ = 3\$6$ is not. H_2O is a syntactically correct chemical formula, but $_2Zz$ is not.

Syntax rules come in two flavors, pertaining to **tokens** and structure. Tokens are the basic elements of the language, such as words, numbers, and chemical elements. One of the problems with $3+ = 3\$6$ is that $\$$ is not a legal token in mathematics (at least as far as I know). Similarly, $_2Zz$ is not legal because there is no element with the abbreviation Zz.

The second type of syntax error pertains to the structure of a statement; that is, the way the tokens are arranged. The statement $3+ = 3$ is illegal because even though + and = are legal tokens, you can't have one right after the other. Similarly, in a chemical formula the subscript comes after the element name, not before.

Exercise 1-1.
Write a well-structured English sentence with invalid tokens in it. Then write another sentence with all valid tokens but with invalid structure.

When you read a sentence in English or a statement in a formal language, you have to figure out what the structure of the sentence is (although in a natural language you do this subconsciously). This process is called **parsing**.

For example, when you hear the sentence, "The penny dropped," you understand that "the penny" is the subject and "dropped" is the predicate. Once you have parsed a sentence, you can figure out what it means, or the semantics of the sentence. Assuming that you know what a penny is and what it means to drop, you will understand the general implication of this sentence.

Although formal and natural languages have many features in common—tokens, structure, syntax, and semantics—there are some differences:

Ambiguity:
> Natural languages are full of ambiguity, which people deal with by using contextual clues and other information. Formal languages are designed to be nearly or completely unambiguous, which means that any statement has exactly one meaning, regardless of context.

Redundancy:
> In order to make up for ambiguity and reduce misunderstandings, natural languages employ lots of redundancy. As a result, they are often verbose. Formal languages are less redundant and more concise.

Literalness:
> Natural languages are full of idiom and metaphor. If I say, "The penny dropped," there is probably no penny and nothing dropping (this idiom means that someone realized something after a period of confusion). Formal languages mean exactly what they say.

People who grow up speaking a natural language (everyone) often have a hard time adjusting to formal languages. In some ways, the difference between formal and natural language is like the difference between poetry and prose, but more so:

Poetry:
> Words are used for their sounds as well as for their meaning, and the whole poem together creates an effect or emotional response. Ambiguity is not only common but often deliberate.

Prose:
> The literal meaning of words is more important, and the structure contributes more meaning. Prose is more amenable to analysis than poetry but still often ambiguous.

Programs:
> The meaning of a computer program is unambiguous and literal, and can be understood entirely by analysis of the tokens and structure.

Here are some suggestions for reading programs (and other formal languages). First, remember that formal languages are much more dense than natural languages, so it takes longer to read them. Also, the structure is very important, so it is usually not a good idea to read from top to bottom, left to right. Instead, learn to parse the program in your head, identifying the tokens and interpreting the structure. Finally, the details matter. Small errors in spelling and punctuation, which you can get away with in natural languages, can make a big difference in a formal language.

The First Program

Traditionally, the first program you write in a new language is called "Hello, World!" because all it does is display the words "Hello, World!". In Python, it looks like this:

```
print 'Hello, World!'
```

This is an example of a **print statement**, which doesn't actually print anything on paper. It displays a value on the screen. In this case, the result is the words

```
Hello, World!
```

The quotation marks in the program mark the beginning and end of the text to be displayed; they don't appear in the result.

In Python 3, the syntax for printing is slightly different:

```
print('Hello, World!')
```

The parentheses indicate that `print` is a function. We'll get to functions in Chapter 3.

For the rest of this book, I'll use the print statement. If you are using Python 3, you will have to translate. But other than that, there are very few differences we have to worry about.

Debugging

It is a good idea to read this book in front of a computer so you can try out the examples as you go. You can run most of the examples in interactive mode, but if you put the code in a script, it is easier to try out variations.

Whenever you are experimenting with a new feature, you should try to make mistakes. For example, in the "Hello, world!" program, what happens if you leave out one of the quotation marks? What if you leave out both? What if you spell `print` wrong?

This kind of experiment helps you remember what you read; it also helps with debugging, because you get to know what the error messages mean. It is better to make mistakes now and on purpose than later and accidentally.

Programming, and especially debugging, sometimes brings out strong emotions. If you are struggling with a difficult bug, you might feel angry, despondent or embarrassed.

There is evidence that people naturally respond to computers as if they were people. When they work well, we think of them as teammates, and when they are obstinate or rude, we respond to them the same way we respond to rude, obstinate people (Reeves and Nass, *The Media Equation: How People Treat Computers, Television, and New Media Like Real People and Places*).

Preparing for these reactions might help you deal with them. One approach is to think of the computer as an employee with certain strengths, like speed and precision, and particular weaknesses, like lack of empathy and inability to grasp the big picture.

Your job is to be a good manager: find ways to take advantage of the strengths and mitigate the weaknesses. And find ways to use your emotions to engage with the problem, without letting your reactions interfere with your ability to work effectively.

Learning to debug can be frustrating, but it is a valuable skill that is useful for many activities beyond programming. At the end of each chapter there is a debugging section, like this one, with my thoughts about debugging. I hope they help!

Glossary

Problem solving:
> The process of formulating a problem, finding a solution, and expressing the solution.

High-level language:
> A programming language like Python that is designed to be easy for humans to read and write.

Low-level language:
> A programming language that is designed to be easy for a computer to execute; also called "machine language" or "assembly language."

Portability:
> A property of a program that can run on more than one kind of computer.

Interpret:
> To execute a program in a high-level language by translating it one line at a time.

Compile:
> To translate a program written in a high-level language into a low-level language all at once, in preparation for later execution.

Source code:
> A program written in a high-level language before being compiled.

Object code:
> The output of the compiler after it translates the program.

Executable:
> Another name for object code that is ready to be executed.

Prompt:
> Characters displayed by the interpreter to indicate that it is ready to take input from the user.

Script:
> A program stored in a file (usually one that will be interpreted).

Interactive mode:
> A way of using the Python interpreter by typing commands and expressions at the prompt.

Script mode:
A way of using the Python interpreter to read and execute statements in a script.

Program:
A set of instructions that specifies a computation.

Algorithm:
A general process for solving a category of problems.

Bug:
An error in a program.

Debugging:
The process of finding and removing any of the three kinds of programming errors.

Syntax:
The structure of a program.

Syntax error:
An error in a program that makes it impossible to parse (and therefore impossible to interpret).

Exception:
An error that is detected while the program is running.

Semantics:
The meaning of a program.

Semantic error:
An error in a program that makes it do something other than what the programmer intended.

Natural language:
Any one of the spoken languages that evolved naturally.

Formal language:
Any one of the languages that people have designed for specific purposes, such as representing mathematical ideas or computer programs; all programming languages are formal languages.

Token:
One of the basic elements of the syntactic structure of a program, analogous to a word in a natural language.

Parse:
To examine a program and analyze the syntactic structure.

Print statement:
An instruction that causes the Python interpreter to display a value on the screen.

Exercises

Exercise 1-2.
Use a web browser to go to the Python website *http://python.org*. This page contains information about Python and links to Python-related pages, and it gives you the ability to search the Python documentation.

Exercise 1-3.
Start the Python interpreter and type help() to start the online help utility. Or you can type help('print') to get information about the print statement.

If this example doesn't work, you may need to install additional Python documentation or set an environment variable; the details depend on your operating system and version of Python.

Exercise 1-4.
Start the Python interpreter and use it as a calculator. Python's syntax for math operations is almost the same as standard mathematical notation. For example, the symbols +, - and / denote addition, subtraction and division, as you would expect. The symbol for multiplication is *.

If you run a 10 kilometer race in 43 minutes 30 seconds, what is your average time per mile? What is your average speed in miles per hour? (Hint: there are 1.61 kilometers in a mile).

$$\frac{1.61K}{M} = \frac{10K}{X}$$

## CHAPTER 2

Variables, Expressions, and Statements

Values and Types

A **value** is one of the basic things a program works with, like a letter or a number. The values we have seen so far are 1, 2, and 'Hello, World!'.

These values belong to different **types**: 2 is an integer, and 'Hello, World!' is a **string**, so-called because it contains a "string" of letters. You (and the interpreter) can identify strings because they are enclosed in quotation marks.

If you are not sure what type a value has, the interpreter can tell you.

```
>>> type('Hello, World!')
<type 'str'>
>>> type(17)
<type 'int'>
```

Not surprisingly, strings belong to the type str and integers belong to the type int. Less obviously, numbers with a decimal point belong to a type called float, because these numbers are represented in a format called **floating-point**.

```
>>> type(3.2)
<type 'float'>
```

What about values like '17' and '3.2'? They look like numbers, but they are in quotation marks like strings.

```
>>> type('17')
<type 'str'>
>>> type('3.2')
<type 'str'>
```

They're strings.

When you type a large integer, you might be tempted to use commas between groups of three digits, as in 1,000,000. This is not a legal integer in Python, but it is legal:

```
>>> 1,000,000
(1, 0, 0)
```

Well, that's not what we expected at all! Python interprets `1,000,000` as a comma-separated sequence of integers. This is the first example we have seen of a semantic error: the code runs without producing an error message, but it doesn't do the "right" thing.

Variables

One of the most powerful features of a programming language is the ability to manipulate **variables**. A variable is a name that refers to a value.

An **assignment statement** creates new variables and gives them values:

```
>>> message = 'And now for something completely different'
>>> n = 17
>>> pi = 3.1415926535897932
```

This example makes three assignments. The first assigns a string to a new variable named `message`; the second gives the integer 17 to `n`; the third assigns the (approximate) value of π to `pi`.

A common way to represent variables on paper is to write the name with an arrow pointing to the variable's value. This kind of figure is called a **state diagram** because it shows what state each of the variables is in (think of it as the variable's state of mind). Figure 2-1 shows the result of the previous example.

Figure 2-1. State diagram.

The type of a variable is the type of the value it refers to.

```
>>> type(message)
<type 'str'>
>>> type(n)
<type 'int'>
>>> type(pi)
<type 'float'>
```

Exercise 2-1.

If you type an integer with a leading zero, you might get a confusing error:

```
>>> zipcode = 02492
                 ^
SyntaxError: invalid token
```

Other numbers seem to work, but the results are bizarre:

```
>>> zipcode = 02132
>>> zipcode
1114
```

Can you figure out what is going on? Hint: display the values 01, 010, 0100 and 01000.

Variable Names and Keywords

Programmers generally choose names for their variables that are meaningful—they document what the variable is used for.

Variable names can be arbitrarily long. They can contain both letters and numbers, but they have to begin with a letter. It is legal to use uppercase letters, but it is a good idea to begin variable names with a lowercase letter (you'll see why later).

The underscore character, _, can appear in a name. It is often used in names with multiple words, such as my_name or airspeed_of_unladen_swallow.

If you give a variable an illegal name, you get a syntax error:

```
>>> 76trombones = 'big parade'
SyntaxError: invalid syntax
>>> more@ = 1000000
SyntaxError: invalid syntax
>>> class = 'Advanced Theoretical Zymurgy'
SyntaxError: invalid syntax
```

76trombones is illegal because it does not begin with a letter. more@ is illegal because it contains an illegal character, @. But what's wrong with class?

It turns out that class is one of Python's **keywords**. The interpreter uses keywords to recognize the structure of the program, and they cannot be used as variable names.

Python 2 has 31 keywords:

and	del	from	not	while
as	elif	global	or	with
assert	else	if	pass	yield
break	except	import	print	
class	exec	in	raise	
continue	finally	is	return	
def	for	lambda	try	

In Python 3, exec is no longer a keyword, but nonlocal is.

You might want to keep this list handy. If the interpreter complains about one of your variable names and you don't know why, see if it is on this list.

Operators and Operands

Operators are special symbols that represent computations like addition and multiplication. The values the operator is applied to are called **operands**.

The operators +, -, *, / and ** perform addition, subtraction, multiplication, division and exponentiation, as in the following examples:

```
20+32    hour-1    hour*60+minute    minute/60    5**2    (5+9)*(15-7)
```

In some other languages, ^ is used for exponentiation, but in Python it is a bitwise operator called XOR. I won't cover bitwise operators in this book, but you can read about them at *http://wiki.python.org/moin/BitwiseOperators*.

In Python 2, the division operator might not do what you expect:

```
>>> minute = 59
>>> minute/60
0
```

The value of `minute` is 59, and in conventional arithmetic 59 divided by 60 is 0.98333, not 0. The reason for the discrepancy is that Python is performing **floor division**. When both of the operands are integers, the result is also an integer; floor division chops off the fraction part, so in this example it rounds down to zero.

In Python 3, the result of this division is a `float`. The new operator `//` performs floor division.

If either of the operands is a floating-point number, Python performs floating-point division, and the result is a `float`:

```
>>> minute/60.0
0.98333333333333328
```

Expressions and Statements

An **expression** is a combination of values, variables, and operators. A value all by itself is considered an expression, and so is a variable, so the following are all legal expressions (assuming that the variable x has been assigned a value):

```
17
x
x + 17
```

A **statement** is a unit of code that the Python interpreter can execute. We have seen two kinds of statement: print and assignment.

Technically an expression is also a statement, but it is probably simpler to think of them as different things. The important difference is that an expression has a value; a statement does not.

Interactive Mode and Script Mode

One of the benefits of working with an interpreted language is that you can test bits of code in interactive mode before you put them in a script. But there are differences between interactive mode and script mode that can be confusing.

For example, if you are using Python as a calculator, you might type

```
>>> miles = 26.2
>>> miles * 1.61
42.182
```

The first line assigns a value to miles, but it has no visible effect. The second line is an expression, so the interpreter evaluates it and displays the result. So we learn that a marathon is about 42 kilometers.

But if you type the same code into a script and run it, you get no output at all. In script mode an expression, all by itself, has no visible effect. Python actually evaluates the expression, but it doesn't display the value unless you tell it to:

```
miles = 26.2
print miles * 1.61
```

This behavior can be confusing at first.

A script usually contains a sequence of statements. If there is more than one statement, the results appear one at a time as the statements execute.

For example, the script

```
print 1
x = 2
print x
```

produces the output

```
1
2
```

The assignment statement produces no output.

Exercise 2-2.
Type the following statements in the Python interpreter to see what they do:

```
5
x = 5
x + 1
```

Now put the same statements into a script and run it. What is the output? Modify the script by transforming each expression into a print statement and then run it again.

Order of Operations

When more than one operator appears in an expression, the order of evaluation depends on the **rules of precedence**. For mathematical operators, Python follows mathematical convention. The acronym **PEMDAS** is a useful way to remember the rules:

- Parentheses have the highest precedence and can be used to force an expression to evaluate in the order you want. Since expressions in parentheses are evaluated first, `2 * (3-1)` is 4, and `(1+1)**(5-2)` is 8. You can also use parentheses to make an expression easier to read, as in `(minute * 100) / 60`, even if it doesn't change the result.

- Exponentiation has the next highest precedence, so `2**1+1` is 3, not 4, and `3*1**3` is 3, not 27.

- Multiplication and Division have the same precedence, which is higher than Addition and Subtraction, which also have the same precedence. So `2*3-1` is 5, not 4, and `6+4/2` is 8, not 5.

- Operators with the same precedence are evaluated from left to right (except exponentiation). So in the expression `degrees / 2 * pi`, the division happens first and the result is multiplied by `pi`. To divide by 2π, you can use parentheses or write `degrees / 2 / pi`.

I don't work very hard to remember rules of precedence for other operators. If I can't tell by looking at the expression, I use parentheses to make it obvious.

String Operations

In general, you can't perform mathematical operations on strings, even if the strings look like numbers, so the following are illegal:

```
'2'-'1'    'eggs'/'easy'    'third'*'a charm'
```

The + operator works with strings, but it might not do what you expect: it performs **concatenation**, which means joining the strings by linking them end-to-end. For example:

```
first = 'throat'
second = 'warbler'
print first + second
```

The output of this program is `throatwarbler`.

The * operator also works on strings; it performs repetition. For example, `'Spam'*3` is `'SpamSpamSpam'`. If one of the operands is a string, the other has to be an integer.

This use of + and * makes sense by analogy with addition and multiplication. Just as 4*3 is equivalent to 4+4+4, we expect 'Spam'*3 to be the same as 'Spam'+'Spam'+'Spam', and it is. On the other hand, there is a significant way in which string concatenation and repetition are different from integer addition and multiplication. Can you think of a property that addition has that string concatenation does not?

Comments

As programs get bigger and more complicated, they get more difficult to read. Formal languages are dense, and it is often difficult to look at a piece of code and figure out what it is doing, or why.

For this reason, it is a good idea to add notes to your programs to explain in natural language what the program is doing. These notes are called **comments**, and they start with the # symbol:

```
# compute the percentage of the hour that has elapsed
percentage = (minute * 100) / 60
```

In this case, the comment appears on a line by itself. You can also put comments at the end of a line:

```
percentage = (minute * 100) / 60     # percentage of an hour
```

Everything from the # to the end of the line is ignored—it has no effect on the program.

Comments are most useful when they document non-obvious features of the code. It is reasonable to assume that the reader can figure out *what* the code does; it is much more useful to explain *why*.

This comment is redundant with the code and useless:

```
v = 5     # assign 5 to v
```

This comment contains useful information that is not in the code:

```
v = 5     # velocity in meters/second.
```

Good variable names can reduce the need for comments, but long names can make complex expressions hard to read, so there is a tradeoff.

Debugging

At this point the syntax error you are most likely to make is an illegal variable name, like class and yield, which are keywords, or odd~job and US$, which contain illegal characters.

If you put a space in a variable name, Python thinks it is two operands without an operator:

```
>>> bad name = 5
SyntaxError: invalid syntax
```

For syntax errors, the error messages don't help much. The most common messages are SyntaxError: invalid syntax and SyntaxError: invalid token, neither of which is very informative.

The runtime error you are most likely to make is a "use before def," that is, trying to use a variable before you have assigned a value. This can happen if you spell a variable name wrong:

```
>>> principal = 327.68
>>> interest = principle * rate
NameError: name 'principle' is not defined
```

Variables names are case sensitive, so LaTeX is not the same as latex.

At this point the most likely cause of a semantic error is the order of operations. For example, to evaluate $\frac{1}{2\pi}$, you might be tempted to write

```
>>> 1.0 / 2.0 * pi
```

But the division happens first, so you would get $\pi / 2$, which is not the same thing! There is no way for Python to know what you meant to write, so in this case you don't get an error message; you just get the wrong answer.

Glossary

Value:
> One of the basic units of data, like a number or string, that a program manipulates.

Type:
> A category of values. The types we have seen so far are integers (type int), floating-point numbers (type float), and strings (type str).

Integer:
> A type that represents whole numbers.

Floating-point:
> A type that represents numbers with fractional parts.

String:
> A type that represents sequences of characters.

Variable:
> A name that refers to a value.

Statement:
> A section of code that represents a command or action. So far, the statements we have seen are assignments and print statements.

Assignment:
   A statement that assigns a value to a variable.

State diagram:
   A graphical representation of a set of variables and the values they refer to.

Keyword:
   A reserved word that is used by the compiler to parse a program; you cannot use keywords like `if`, `def`, and `while` as variable names.

Operator:
   A special symbol that represents a simple computation like addition, multiplication, or string concatenation.

Operand:
   One of the values on which an operator operates.

Floor division:
   The operation that divides two numbers and chops off the fraction part.

Expression:
   A combination of variables, operators, and values that represents a single result value.

Evaluate:
   To simplify an expression by performing the operations in order to yield a single value.

Rules of precedence:
   The set of rules governing the order in which expressions involving multiple operators and operands are evaluated.

Concatenate:
   To join two operands end-to-end.

Comment:
   Information in a program that is meant for other programmers (or anyone reading the source code) and has no effect on the execution of the program.

Exercises

Exercise 2-3.
Assume that we execute the following assignment statements:

```
width = 17
height = 12.0
delimiter = '.'
```

For each of the following expressions, write the value of the expression and the type (of the value of the expression).

1. `width/2`
2. `width/2.0`
3. `height/3`
4. `1 + 2 * 5`
5. `delimiter * 5`

Use the Python interpreter to check your answers.

Exercise 2-4.

Practice using the Python interpreter as a calculator:

1. The volume of a sphere with radius r is $\frac{4}{3}\pi r^3$. What is the volume of a sphere with radius 5? Hint: 392.7 is wrong!

2. Suppose the cover price of a book is $24.95, but bookstores get a 40% discount. Shipping costs $3 for the first copy and 75 cents for each additional copy. What is the total wholesale cost for 60 copies?

3. If I leave my house at 6:52 am and run 1 mile at an easy pace (8:15 per mile), then 3 miles at tempo (7:12 per mile) and 1 mile at easy pace again, what time do I get home for breakfast?

## CHAPTER 3

Functions

Function Calls

In the context of programming, a **function** is a named sequence of statements that performs a computation. When you define a function, you specify the name and the sequence of statements. Later, you can "call" the function by name. We have already seen one example of a **function call**:

```
>>> type(32)
<type 'int'>
```

The name of the function is type. The expression in parentheses is called the **argument** of the function. The result, for this function, is the type of the argument.

It is common to say that a function "takes" an argument and "returns" a result. The result is called the **return value**.

Type Conversion Functions

Python provides built-in functions that convert values from one type to another. The int function takes any value and converts it to an integer, if it can, or complains otherwise:

```
>>> int('32')
32
>>> int('Hello')
ValueError: invalid literal for int(): Hello
```

int can convert floating-point values to integers, but it doesn't round off; it chops off the fraction part:

```
>>> int(3.99999)
3
>>> int(-2.3)
-2
```

`float` converts integers and strings to floating-point numbers:

```
>>> float(32)
32.0
>>> float('3.14159')
3.14159
```

Finally, `str` converts its argument to a string:

```
>>> str(32)
'32'
>>> str(3.14159)
'3.14159'
```

Math Functions

Python has a math module that provides most of the familiar mathematical functions. A **module** is a file that contains a collection of related functions.

Before we can use the module, we have to import it:

```
>>> import math
```

This statement creates a **module object** named math. If you print the module object, you get some information about it:

```
>>> print math
<module 'math' (built-in)>
```

The module object contains the functions and variables defined in the module. To access one of the functions, you have to specify the name of the module and the name of the function, separated by a dot (also known as a period). This format is called **dot notation**.

```
>>> ratio = signal_power / noise_power
>>> decibels = 10 * math.log10(ratio)

>>> radians = 0.7
>>> height = math.sin(radians)
```

The first example uses `log10` to compute a signal-to-noise ratio in decibels (assuming that `signal_power` and `noise_power` are defined). The math module also provides `log`, which computes logarithms base e.

The second example finds the sine of `radians`. The name of the variable is a hint that `sin` and the other trigonometric functions (`cos`, `tan`, etc.) take arguments in radians. To convert from degrees to radians, divide by 360 and multiply by 2π:

```
>>> degrees = 45
>>> radians = degrees / 360.0 * 2 * math.pi
>>> math.sin(radians)
0.707106781187
```

The expression `math.pi` gets the variable `pi` from the math module. The value of this variable is an approximation of π, accurate to about 15 digits.

If you know your trigonometry, you can check the previous result by comparing it to the square root of two divided by two:

```
>>> math.sqrt(2) / 2.0
0.707106781187
```

Composition

So far, we have looked at the elements of a program—variables, expressions, and statements—in isolation, without talking about how to combine them.

One of the most useful features of programming languages is their ability to take small building blocks and **compose** them. For example, the argument of a function can be any kind of expression, including arithmetic operators:

```
x = math.sin(degrees / 360.0 * 2 * math.pi)
```

And even function calls:

```
x = math.exp(math.log(x+1))
```

Almost anywhere you can put a value, you can put an arbitrary expression, with one exception: the left side of an assignment statement has to be a variable name. Any other expression on the left side is a syntax error (we will see exceptions to this rule later).

```
>>> minutes = hours * 60          # right
>>> hours * 60 = minutes          # wrong!
SyntaxError: can't assign to operator
```

Adding New Functions

So far, we have only been using the functions that come with Python, but it is also possible to add new functions. A **function definition** specifies the name of a new function and the sequence of statements that execute when the function is called.

Here is an example:

```
def print_lyrics():
    print "I'm a lumberjack, and I'm okay."
    print "I sleep all night and I work all day."
```

`def` is a keyword that indicates that this is a function definition. The name of the function is `print_lyrics`. The rules for function names are the same as for variable names: letters, numbers and some punctuation marks are legal, but the first character can't be a number. You can't use a keyword as the name of a function, and you should avoid having a variable and a function with the same name.

The empty parentheses after the name indicate that this function doesn't take any arguments.

The first line of the function definition is called the **header**; the rest is called the **body**. The header has to end with a colon and the body has to be indented. By convention, the indentation is always four spaces; see "Debugging" (page 33). The body can contain any number of statements.

The strings in the print statements are enclosed in double quotes. Single quotes and double quotes do the same thing; most people use single quotes except in cases like this where a single quote (which is also an apostrophe) appears in the string.

If you type a function definition in interactive mode, the interpreter prints ellipses (...) to let you know that the definition isn't complete:

```
>>> def print_lyrics():
...     print "I'm a lumberjack, and I'm okay."
...     print "I sleep all night and I work all day."
...
```

To end the function, you have to enter an empty line (this is not necessary in a script).

Defining a function creates a variable with the same name.

```
>>> print print_lyrics
<function print_lyrics at 0xb7e99e9c>
>>> type(print_lyrics)
<type 'function'>
```

The value of `print_lyrics` is a **function object**, which has type `'function'`.

The syntax for calling the new function is the same as for built-in functions:

```
>>> print_lyrics()
I'm a lumberjack, and I'm okay.
I sleep all night and I work all day.
```

Once you have defined a function, you can use it inside another function. For example, to repeat the previous refrain, we could write a function called `repeat_lyrics`:

```
def repeat_lyrics():
    print_lyrics()
    print_lyrics()
```

And then call `repeat_lyrics`:

```
>>> repeat_lyrics()
I'm a lumberjack, and I'm okay.
I sleep all night and I work all day.
I'm a lumberjack, and I'm okay.
I sleep all night and I work all day.
```

But that's not really how the song goes.

Definitions and Uses

Pulling together the code fragments from the previous section, the whole program looks like this:

```
def print_lyrics():
    print "I'm a lumberjack, and I'm okay."
    print "I sleep all night and I work all day."

def repeat_lyrics():
    print_lyrics()
    print_lyrics()

repeat_lyrics()
```

This program contains two function definitions: `print_lyrics` and `repeat_lyrics`. Function definitions get executed just like other statements, but the result creates function objects. The statements inside the function do not get executed until the function is called, and the function definition generates no output.

As you might expect, you have to create a function before you can execute it. In other words, the function definition has to be executed before the function is called the first time.

Exercise 3-1.
Move the last line of this program to the top, so the function call appears before the definitions. Run the program and see what error message you get.

Exercise 3-2.
Move the function call back to the bottom and move the definition of `print_lyrics` after the definition of `repeat_lyrics`. What happens when you run this program?

Flow of Execution

In order to ensure that a function is defined before its first use, you have to know the order in which statements are executed, which is called the **flow of execution**.

Execution always begins at the first statement of the program. Statements are executed one at a time, in order, from top to bottom.

Function definitions do not alter the flow of execution of the program, but remember that statements inside the function are not executed until the function is called.

A function call is like a detour in the flow of execution. Instead of going to the next statement, the flow jumps to the body of the function, executes all the statements there, and then comes back to pick up where it left off.

That sounds simple enough, until you remember that one function can call another. While in the middle of one function, the program might have to execute the statements in another function. But while executing that new function, the program might have to execute yet another function!

Fortunately, Python is good at keeping track of where it is, so each time a function completes, the program picks up where it left off in the function that called it. When it gets to the end of the program, it terminates.

What's the moral of this sordid tale? When you read a program, you don't always want to read from top to bottom. Sometimes it makes more sense if you follow the flow of execution.

Parameters and Arguments

Some of the built-in functions we have seen require arguments. For example, when you call math.sin you pass a number as an argument. Some functions take more than one argument: math.pow takes two, the base and the exponent.

Inside the function, the arguments are assigned to variables called **parameters**. Here is an example of a user-defined function that takes an argument:

```
def print_twice(bruce):
    print bruce
    print bruce
```

This function assigns the argument to a parameter named bruce. When the function is called, it prints the value of the parameter (whatever it is) twice.

This function works with any value that can be printed.

```
>>> print_twice('Spam')
Spam
Spam
>>> print_twice(17)
17
17
>>> print_twice(math.pi)
3.14159265359
3.14159265359
```

The same rules of composition that apply to built-in functions also apply to user-defined functions, so we can use any kind of expression as an argument for print_twice:

```
>>> print_twice('Spam '*4)
Spam Spam Spam Spam
Spam Spam Spam Spam
>>> print_twice(math.cos(math.pi))
-1.0
-1.0
```

The argument is evaluated before the function is called, so in the examples the expressions `'Spam '*4` and `math.cos(math.pi)` are only evaluated once.

You can also use a variable as an argument:

```
>>> michael = 'Eric, the half a bee.'
>>> print_twice(michael)
Eric, the half a bee.
Eric, the half a bee.
```

The name of the variable we pass as an argument (`michael`) has nothing to do with the name of the parameter (`bruce`). It doesn't matter what the value was called back home (in the caller); here in `print_twice`, we call everybody `bruce`.

Variables and Parameters Are Local

When you create a variable inside a function, it is **local**, which means that it only exists inside the function. For example:

```
def cat_twice(part1, part2):
    cat = part1 + part2
    print_twice(cat)
```

This function takes two arguments, concatenates them, and prints the result twice. Here is an example that uses it:

```
>>> line1 = 'Bing tiddle '
>>> line2 = 'tiddle bang.'
>>> cat_twice(line1, line2)
Bing tiddle tiddle bang.
Bing tiddle tiddle bang.
```

When `cat_twice` terminates, the variable `cat` is destroyed. If we try to print it, we get an exception:

```
>>> print cat
NameError: name 'cat' is not defined
```

Parameters are also local. For example, outside `print_twice`, there is no such thing as `bruce`.

Stack Diagrams

To keep track of which variables can be used where, it is sometimes useful to draw a **stack diagram**. Like state diagrams, stack diagrams show the value of each variable, but they also show the function each variable belongs to.

Each function is represented by a **frame**. A frame is a box with the name of a function beside it and the parameters and variables of the function inside it. The stack diagram for the previous example is shown in Figure 3-1.

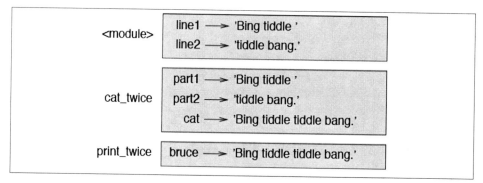

Figure 3-1. Stack diagram.

The frames are arranged in a stack that indicates which function called which, and so on. In this example, print_twice was called by cat_twice, and cat_twice was called by __main__, which is a special name for the topmost frame. When you create a variable outside of any function, it belongs to __main__.

Each parameter refers to the same value as its corresponding argument. So, part1 has the same value as line1, part2 has the same value as line2, and bruce has the same value as cat.

If an error occurs during a function call, Python prints the name of the function, and the name of the function that called it, and the name of the function that called *that*, all the way back to __main__.

For example, if you try to access cat from within print_twice, you get a NameError:

```
Traceback (innermost last):
  File "test.py", line 13, in __main__
    cat_twice(line1, line2)
  File "test.py", line 5, in cat_twice
    print_twice(cat)
  File "test.py", line 9, in print_twice
    print cat
NameError: name 'cat' is not defined
```

This list of functions is called a **traceback**. It tells you what program file the error occurred in, and what line, and what functions were executing at the time. It also shows the line of code that caused the error.

The order of the functions in the traceback is the same as the order of the frames in the stack diagram. The function that is currently running is listed at the bottom.

Fruitful Functions and Void Functions

Some of the functions we are using, such as the math functions, yield results; for lack of a better name, I call them **fruitful functions**. Other functions, like `print_twice`, perform an action but don't return a value. They are called **void functions**.

When you call a fruitful function, you almost always want to do something with the result; for example, you might assign it to a variable or use it as part of an expression:

```
x = math.cos(radians)
golden = (math.sqrt(5) + 1) / 2
```

When you call a function in interactive mode, Python displays the result:

```
>>> math.sqrt(5)
2.2360679774997898
```

But in a script, if you call a fruitful function all by itself, the return value is lost forever!

```
math.sqrt(5)
```

This script computes the square root of 5, but since it doesn't store or display the result, it is not very useful.

Void functions might display something on the screen or have some other effect, but they don't have a return value. If you try to assign the result to a variable, you get a special value called None.

```
>>> result = print_twice('Bing')
Bing
Bing
>>> print result
None
```

The value None is not the same as the string 'None'. It is a special value that has its own type:

```
>>> print type(None)
<type 'NoneType'>
```

The functions we have written so far are all void. We will start writing fruitful functions in a few chapters.

Why Functions?

It may not be clear why it is worth the trouble to divide a program into functions. There are several reasons:

- Creating a new function gives you an opportunity to name a group of statements, which makes your program easier to read and debug.

- Functions can make a program smaller by eliminating repetitive code. Later, if you make a change, you only have to make it in one place.

- Dividing a long program into functions allows you to debug the parts one at a time and then assemble them into a working whole.

- Well-designed functions are often useful for many programs. Once you write and debug one, you can reuse it.

Importing with from

Python provides two ways to import modules; we have already seen one:

```
>>> import math
>>> print math
<module 'math' (built-in)>
>>> print math.pi
3.14159265359
```

If you import math, you get a module object named math. The module object contains constants like pi and functions like sin and exp.

But if you try to access pi directly, you get an error.

```
>>> print pi
Traceback (most recent call last):
  File "<stdin>", line 1, in <module>
NameError: name 'pi' is not defined
```

As an alternative, you can import an object from a module like this:

```
>>> from math import pi
```

Now you can access pi directly, without dot notation.

```
>>> print pi
3.14159265359
```

Or you can use the star operator to import *everything* from the module:

```
>>> from math import *
>>> cos(pi)
-1.0
```

The advantage of importing everything from the math module is that your code can be more concise. The disadvantage is that there might be conflicts between names defined in different modules, or between a name from a module and one of your variables.

Debugging

If you are using a text editor to write your scripts, you might run into problems with spaces and tabs. The best way to avoid these problems is to use spaces exclusively (no tabs). Most text editors that know about Python do this by default, but some don't.

Tabs and spaces are usually invisible, which makes them hard to debug, so try to find an editor that manages indentation for you.

Also, don't forget to save your program before you run it. Some development environments do this automatically, but some don't. In that case the program you are looking at in the text editor is not the same as the program you are running.

Debugging can take a long time if you keep running the same, incorrect, program over and over!

Make sure that the code you are looking at is the code you are running. If you're not sure, put something like `print 'hello'` at the beginning of the program and run it again. If you don't see `hello`, you're not running the right program!

Glossary

Function:
> A named sequence of statements that performs some useful operation. Functions may or may not take arguments and may or may not produce a result.

Function definition:
> A statement that creates a new function, specifying its name, parameters, and the statements it executes.

Function object:
> A value created by a function definition. The name of the function is a variable that refers to a function object.

Header:
> The first line of a function definition.

Body:
> The sequence of statements inside a function definition.

Parameter:
> A name used inside a function to refer to the value passed as an argument.

Function call:

A statement that executes a function. It consists of the function name followed by an argument list.

Argument:

A value provided to a function when the function is called. This value is assigned to the corresponding parameter in the function.

Local variable:

A variable defined inside a function. A local variable can only be used inside its function.

Return value:

The result of a function. If a function call is used as an expression, the return value is the value of the expression.

Fruitful function:

A function that returns a value.

Void function:

A function that doesn't return a value.

Module:

A file that contains a collection of related functions and other definitions.

Import statement:

A statement that reads a module file and creates a module object.

Module object:

A value created by an `import` statement that provides access to the values defined in a module.

Dot notation:

The syntax for calling a function in another module by specifying the module name followed by a dot (period) and the function name.

Composition:

Using an expression as part of a larger expression, or a statement as part of a larger statement.

Flow of execution:

The order in which statements are executed during a program run.

Stack diagram:

A graphical representation of a stack of functions, their variables, and the values they refer to.

Frame:
A box in a stack diagram that represents a function call. It contains the local variables and parameters of the function.

Traceback:
A list of the functions that are executing, printed when an exception occurs.

Exercises

Exercise 3-3.
Python provides a built-in function called `len` that returns the length of a string, so the value of `len('allen')` is 5.

Write a function named `right_justify` that takes a string named `s` as a parameter and prints the string with enough leading spaces so that the last letter of the string is in column 70 of the display.

```
>>> right_justify('allen')
                                                                 allen
```

Exercise 3-4.
A function object is a value you can assign to a variable or pass as an argument. For example, `do_twice` is a function that takes a function object as an argument and calls it twice:

```
def do_twice(f):
    f()
    f()
```

Here's an example that uses `do_twice` to call a function named `print_spam` twice.

```
def print_spam():
    print 'spam'

do_twice(print_spam)
```

1. Type this example into a script and test it.

2. Modify `do_twice` so that it takes two arguments, a function object and a value, and calls the function twice, passing the value as an argument.

3. Write a more general version of `print_spam`, called `print_twice`, that takes a string as a parameter and prints it twice.

4. Use the modified version of `do_twice` to call `print_twice` twice, passing `'spam'` as an argument.

5. Define a new function called do_four that takes a function object and a value and calls the function four times, passing the value as a parameter. There should be only two statements in the body of this function, not four.

Solution: *http://thinkpython.com/code/do_four.py*.

Exercise 3-5.

This exercise can be done using only the statements and other features we have learned so far.

1. Write a function that draws a grid like the following:

```
+ - - - - + - - - - +
|         |         |
|         |         |
|         |         |
|         |         |
+ - - - - + - - - - +
|         |         |
|         |         |
|         |         |
|         |         |
+ - - - - + - - - - +
```

Hint: to print more than one value on a line, you can print a comma-separated sequence:

```
print '+', '-'
```

If the sequence ends with a comma, Python leaves the line unfinished, so the value printed next appears on the same line.

```
print '+',
print '-'
```

The output of these statements is '+ -'.

A print statement all by itself ends the current line and goes to the next line.

2. Write a function that draws a similar grid with four rows and four columns.

Solution: *http://thinkpython.com/code/grid.py*. Credit: This exercise is based on an exercise in Oualline, *Practical C Programming, Third Edition*, O'Reilly Media, 1997.

## CHAPTER 4

Case Study: Interface Design

Code examples from this chapter are available from *http://thinkpython.com/code/ polygon.py*.

TurtleWorld

To accompany this book, I have written a package called Swampy. You can download Swampy from *http://thinkpython.com/swampy*; follow the instructions there to install Swampy on your system.

A **package** is a collection of modules; one of the modules in Swampy is TurtleWorld, which provides a set of functions for drawing lines by steering turtles around the screen.

If Swampy is installed as a package on your system, you can import TurtleWorld like this:

```
from swampy.TurtleWorld import *
```

If you downloaded the Swampy modules but did not install them as a package, you can either work in the directory that contains the Swampy files, or add that directory to Python's search path. Then you can import TurtleWorld like this:

```
from TurtleWorld import *
```

The details of the installation process and setting Python's search path depend on your system, so rather than include those details here, I will try to maintain current information for several systems at *http://thinkpython.com/swampy*

Create a file named mypolygon.py and type in the following code:

```
from swampy.TurtleWorld import *

world = TurtleWorld()
```

```
bob = Turtle()
print bob

wait_for_user()
```

The first line imports everything from the `TurtleWorld` module in the `swampy` package.

The next lines create a TurtleWorld assigned to `world` and a Turtle assigned to `bob`. Printing `bob` yields something like:

```
<TurtleWorld.Turtle instance at 0xb7bfbf4c>
```

This means that `bob` refers to an **instance** of a Turtle as defined in module `Turtle World`. In this context, "instance" means a member of a set; this Turtle is one of the set of possible Turtles.

`wait_for_user` tells TurtleWorld to wait for the user to do something, although in this case there's not much for the user to do except close the window.

TurtleWorld provides several turtle-steering functions: `fd` and `bk` for forward and backward, and `lt` and `rt` for left and right turns. Also, each Turtle is holding a pen, which is either down or up; if the pen is down, the Turtle leaves a trail when it moves. The functions `pu` and `pd` stand for "pen up" and "pen down."

To draw a right angle, add these lines to the program (after creating `bob` and before calling `wait_for_user`):

```
fd(bob, 100)
lt(bob)
fd(bob, 100)
```

The first line tells `bob` to take 100 steps forward. The second line tells him to turn left.

When you run this program, you should see `bob` move east and then north, leaving two line segments behind.

Now modify the program to draw a square. Don't go on until you've got it working!

Simple Repetition

Chances are you wrote something like this (leaving out the code that creates TurtleWorld and waits for the user):

```
fd(bob, 100)
lt(bob)

fd(bob, 100)
lt(bob)
```

```
fd(bob, 100)
lt(bob)

fd(bob, 100)
```

We can do the same thing more concisely with a `for` statement. Add this example to `mypolygon.py` and run it again:

```
for i in range(4):
    print 'Hello!'
```

You should see something like this:

```
Hello!
Hello!
Hello!
Hello!
```

This is the simplest use of the `for` statement; we will see more later. But that should be enough to let you rewrite your square-drawing program. Don't go on until you do.

Here is a `for` statement that draws a square:

```
for i in range(4):
    fd(bob, 100)
    lt(bob)
```

The syntax of a `for` statement is similar to a function definition. It has a header that ends with a colon and an indented body. The body can contain any number of statements.

A `for` statement is sometimes called a **loop** because the flow of execution runs through the body and then loops back to the top. In this case, it runs the body four times.

This version is actually a little different from the previous square-drawing code because it makes another turn after drawing the last side of the square. The extra turn takes a little more time, but it simplifies the code if we do the same thing every time through the loop. This version also has the effect of leaving the turtle back in the starting position, facing in the starting direction.

Exercises

The following is a series of exercises using TurtleWorld. They are meant to be fun, but they have a point, too. While you are working on them, think about what the point is.

The following sections have solutions to the exercises, so don't look until you have finished (or at least tried).

1. Write a function called square that takes a parameter named t, which is a turtle. It should use the turtle to draw a square.

   Write a function call that passes bob as an argument to square, and then run the program again.

2. Add another parameter, named length, to square. Modify the body so length of the sides is length, and then modify the function call to provide a second argument. Run the program again. Test your program with a range of values for length.

3. The functions lt and rt make 90-degree turns by default, but you can provide a second argument that specifies the number of degrees. For example, lt(bob, 45) turns bob 45 degrees to the left.

   Make a copy of square and change the name to polygon. Add another parameter named n and modify the body so it draws an n-sided regular polygon. Hint: The exterior angles of an n-sided regular polygon are *360/n* degrees.

4. Write a function called circle that takes a turtle, t, and radius, r, as parameters and that draws an approximate circle by invoking polygon with an appropriate length and number of sides. Test your function with a range of values of r.

   Hint: figure out the circumference of the circle and make sure that length * n = circumference.

   Another hint: if bob is too slow for you, you can speed him up by changing bob.delay, which is the time between moves, in seconds. bob.delay = 0.01 ought to get him moving.

5. Make a more general version of circle called arc that takes an additional parameter angle, which determines what fraction of a circle to draw. angle is in units of degrees, so when angle=360, arc should draw a complete circle.

Encapsulation

The first exercise asks you to put your square-drawing code into a function definition and then call the function, passing the turtle as a parameter. Here is a solution:

```
def square(t):
    for i in range(4):
        fd(t, 100)
        lt(t)

square(bob)
```

The innermost statements, fd and lt are indented twice to show that they are inside the for loop, which is inside the function definition. The next line, square(bob), is flush with the left margin, so that is the end of both the for loop and the function definition.

Inside the function, t refers to the same turtle bob refers to, so lt(t) has the same effect as lt(bob). So why not call the parameter bob? The idea is that t can be any turtle, not just bob, so you could create a second turtle and pass it as an argument to square:

```
ray = Turtle()
square(ray)
```

Wrapping a piece of code up in a function is called **encapsulation**. One of the benefits of encapsulation is that it attaches a name to the code, which serves as a kind of documentation. Another advantage is that if you reuse the code, it is more concise to call a function twice than to copy and paste the body!

Generalization

The next step is to add a length parameter to square. Here is a solution:

```
def square(t, length):
    for i in range(4):
        fd(t, length)
        lt(t)

square(bob, 100)
```

Adding a parameter to a function is called **generalization** because it makes the function more general: in the previous version, the square is always the same size; in this version it can be any size.

The next step is also a generalization. Instead of drawing squares, polygon draws regular polygons with any number of sides. Here is a solution:

```
def polygon(t, n, length):
    angle = 360.0 / n
    for i in range(n):
        fd(t, length)
        lt(t, angle)

polygon(bob, 7, 70)
```

This draws a 7-sided polygon with side length 70. If you have more than a few numeric arguments, it is easy to forget what they are, or what order they should be in. It is legal, and sometimes helpful, to include the names of the parameters in the argument list:

```
polygon(bob, n=7, length=70)
```

These are called **keyword arguments** because they include the parameter names as "keywords" (not to be confused with Python keywords like while and def).

This syntax makes the program more readable. It is also a reminder about how arguments and parameters work: when you call a function, the arguments are assigned to the parameters.

Interface Design

The next step is to write `circle`, which takes a radius, r, as a parameter. Here is a simple solution that uses `polygon` to draw a 50-sided polygon:

```
def circle(t, r):
    circumference = 2 * math.pi * r
    n = 50
    length = circumference / n
    polygon(t, n, length)
```

The first line computes the circumference of a circle with radius r using the formula $2\pi r$. Since we use `math.pi`, we have to import `math`. By convention, `import` statements are usually at the beginning of the script.

n is the number of line segments in our approximation of a circle, so `length` is the length of each segment. Thus, `polygon` draws a 50-sided polygon that approximates a circle with radius r.

One limitation of this solution is that n is a constant, which means that for very big circles, the line segments are too long, and for small circles, we waste time drawing very small segments. One solution would be to generalize the function by taking n as a parameter. This would give the user (whoever calls `circle`) more control, but the interface would be less clean.

The **interface** of a function is a summary of how it is used: what are the parameters? What does the function do? And what is the return value? An interface is "clean" if it is "as simple as possible, but not simpler. (Einstein)"

In this example, r belongs in the interface because it specifies the circle to be drawn. n is less appropriate because it pertains to the details of *how* the circle should be rendered.

Rather than clutter up the interface, it is better to choose an appropriate value of n depending on `circumference`:

```
def circle(t, r):
    circumference = 2 * math.pi * r
    n = int(circumference / 3) + 1
    length = circumference / n
    polygon(t, n, length)
```

Now the number of segments is (approximately) `circumference`/3, so the length of each segment is (approximately) 3, which is small enough that the circles look good, but big enough to be efficient, and appropriate for any size circle.

Refactoring

When I wrote `circle`, I was able to reuse `polygon` because a many-sided polygon is a good approximation of a circle. But `arc` is not as cooperative; we can't use `polygon` or `circle` to draw an arc.

One alternative is to start with a copy of `polygon` and transform it into `arc`. The result might look like this:

```
def arc(t, r, angle):
    arc_length = 2 * math.pi * r * angle / 360
    n = int(arc_length / 3) + 1
    step_length = arc_length / n
    step_angle = float(angle) / n

    for i in range(n):
        fd(t, step_length)
        lt(t, step_angle)
```

The second half of this function looks like `polygon`, but we can't reuse `polygon` without changing the interface. We could generalize `polygon` to take an angle as a third argument, but then `polygon` would no longer be an appropriate name! Instead, let's call the more general function `polyline`:

```
def polyline(t, n, length, angle):
    for i in range(n):
        fd(t, length)
        lt(t, angle)
```

Now we can rewrite `polygon` and `arc` to use `polyline`:

```
def polygon(t, n, length):
    angle = 360.0 / n
    polyline(t, n, length, angle)

def arc(t, r, angle):
    arc_length = 2 * math.pi * r * angle / 360
    n = int(arc_length / 3) + 1
    step_length = arc_length / n
    step_angle = float(angle) / n
    polyline(t, n, step_length, step_angle)
```

Finally, we can rewrite `circle` to use `arc`:

```
def circle(t, r):
    arc(t, r, 360)
```

This process—rearranging a program to improve function interfaces and facilitate code reuse—is called **refactoring**. In this case, we noticed that there was similar code in `arc` and `polygon`, so we "factored it out" into `polyline`.

If we had planned ahead, we might have written polyline first and avoided refactoring, but often you don't know enough at the beginning of a project to design all the interfaces. Once you start coding, you understand the problem better. Sometimes refactoring is a sign that you have learned something.

A Development Plan

A **development plan** is a process for writing programs. The process we used in this case study is "encapsulation and generalization." The steps of this process are:

1. Start by writing a small program with no function definitions.

2. Once you get the program working, encapsulate it in a function and give it a name.

3. Generalize the function by adding appropriate parameters.

4. Repeat steps 1–3 until you have a set of working functions. Copy and paste working code to avoid retyping (and re-debugging).

5. Look for opportunities to improve the program by refactoring. For example, if you have similar code in several places, consider factoring it into an appropriately general function.

This process has some drawbacks—we will see alternatives later—but it can be useful if you don't know ahead of time how to divide the program into functions. This approach lets you design as you go along.

Docstring

A **docstring** is a string at the beginning of a function that explains the interface ("doc" is short for "documentation"). Here is an example:

```
def polyline(t, n, length, angle):
    """Draws n line segments with the given length and
    angle (in degrees) between them.  t is a turtle.
    """
    for i in range(n):
        fd(t, length)
        lt(t, angle)
```

This docstring is a triple-quoted string, also known as a multiline string because the triple quotes allow the string to span more than one line.

It is terse, but it contains the essential information someone would need to use this function. It explains concisely what the function does (without getting into the details of how it does it). It explains what effect each parameter has on the behavior of the function and what type each parameter should be (if it is not obvious).

Writing this kind of documentation is an important part of interface design. A well-designed interface should be simple to explain; if you are having a hard time explaining one of your functions, that might be a sign that the interface could be improved.

Debugging

An interface is like a contract between a function and a caller. The caller agrees to provide certain parameters and the function agrees to do certain work.

For example, polyline requires four arguments: t has to be a Turtle; n is the number of line segments, so it has to be an integer; length should be a positive number; and angle has to be a number, which is understood to be in degrees.

These requirements are called **preconditions** because they are supposed to be true before the function starts executing. Conversely, conditions at the end of the function are **postconditions**. Postconditions include the intended effect of the function (like drawing line segments) and any side effects (like moving the Turtle or making other changes in the World).

Preconditions are the responsibility of the caller. If the caller violates a (properly documented!) precondition and the function doesn't work correctly, the bug is in the caller, not the function.

Glossary

Instance:
> A member of a set. The TurtleWorld in this chapter is a member of the set of TurtleWorlds.

Loop:
> A part of a program that can execute repeatedly.

Encapsulation:
> The process of transforming a sequence of statements into a function definition.

Generalization:
> The process of replacing something unnecessarily specific (like a number) with something appropriately general (like a variable or parameter).

Keyword argument:
> An argument that includes the name of the parameter as a "keyword."

Interface:
> A description of how to use a function, including the name and descriptions of the arguments and return value.

Figure 4-1. Turtle flowers.

Refactoring:
  The process of modifying a working program to improve function interfaces and other qualities of the code.

Development plan:
  A process for writing programs.

Docstring:
  A string that appears in a function definition to document the function's interface.

Precondition:
  A requirement that should be satisfied by the caller before a function starts.

Postcondition:
  A requirement that should be satisfied by the function before it ends.

Exercises

Exercise 4-1.
Download the code in this chapter from *http://thinkpython.com/code/polygon.py*.

1. Write appropriate docstrings for `polygon`, `arc` and `circle`.

2. Draw a stack diagram that shows the state of the program while executing `circle(bob, radius)`. You can do the arithmetic by hand or add `print` statements to the code.

3. The version of `arc` in "Refactoring" (page 43) is not very accurate because the linear approximation of the circle is always outside the true circle. As a result, the turtle ends up a few units away from the correct destination. My solution shows a way to reduce the effect of this error. Read the code and see if it makes sense to you. If you draw a diagram, you might see how it works.

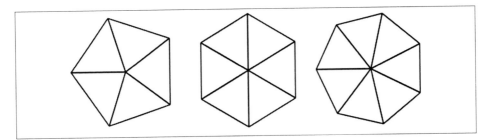

Figure 4-2. Turtle pies.

Exercise 4-2.
Write an appropriately general set of functions that can draw flowers as in Figure 4-1.

Solution: *http://thinkpython.com/code/flower.py*, also requires *http://thinkpython.com/code/polygon.py*.

Exercise 4-3.
Write an appropriately general set of functions that can draw shapes as in Figure 4-2.

Solution: *http://thinkpython.com/code/pie.py*.

Exercise 4-4.
The letters of the alphabet can be constructed from a moderate number of basic elements, like vertical and horizontal lines and a few curves. Design a font that can be drawn with a minimal number of basic elements and then write functions that draw letters of the alphabet.

You should write one function for each letter, with names draw_a, draw_b, etc., and put your functions in a file named letters.py. You can download a "turtle typewriter" from *http://thinkpython.com/code/typewriter.py* to help you test your code.

Solution: *http://thinkpython.com/code/letters.py*, also requires *http://thinkpython.com/code/polygon.py*.

Exercise 4-5.
Read about spirals at *http://en.wikipedia.org/wiki/Spiral*; then write a program that draws an Archimedian spiral (or one of the other kinds). Solution: *http://thinkpython.com/code/spiral.py*.

## CHAPTER 5

Conditionals and Recursion

Modulus Operator

The **modulus operator** works on integers and yields the remainder when the first operand is divided by the second. In Python, the modulus operator is a percent sign (%). The syntax is the same as for other operators:

```
>>> quotient = 7 / 3
>>> print quotient
2
>>> remainder = 7 % 3
>>> print remainder
1
```

So 7 divided by 3 is 2 with 1 left over.

The modulus operator turns out to be surprisingly useful. For example, you can check whether one number is divisible by another—if x % y is zero, then x is divisible by y.

Also, you can extract the right-most digit or digits from a number. For example, x % 10 yields the right-most digit of x (in base 10). Similarly x % 100 yields the last two digits.

Boolean Expressions

A **Boolean expression** is an expression that is either true or false. The following examples use the operator ==, which compares two operands and produces True if they are equal and False otherwise:

```
>>> 5 == 5
True
>>> 5 == 6
False
```

`True` and `False` are special values that belong to the type `bool`; they are not strings:

```
>>> type(True)
<type 'bool'>
>>> type(False)
<type 'bool'>
```

The `==` operator is one of the **relational operators**; the others are:

```
x != y      # x is not equal to y
x > y       # x is greater than y
x < y       # x is less than y
x >= y      # x is greater than or equal to y
x <= y      # x is less than or equal to y
```

Although these operations are probably familiar to you, the Python symbols are different from the mathematical symbols. A common error is to use a single equal sign (=) instead of a double equal sign (==). Remember that = is an assignment operator and == is a relational operator. There is no such thing as =< or =>.

Logical Operators

There are three **logical operators**: `and`, `or`, and `not`. The semantics (meaning) of these operators is similar to their meaning in English. For example, `x > 0 and x < 10` is true only if x is greater than 0 *and* less than 10.

`n%2 == 0 or n%3 == 0` is true if *either* of the conditions is true, that is, if the number is divisible by 2 *or* 3.

Finally, the `not` operator negates a boolean expression, so `not (x > y)` is true if x > y is false, that is, if x is less than or equal to y.

Strictly speaking, the operands of the logical operators should be boolean expressions, but Python is not very strict. Any nonzero number is interpreted as "true."

```
>>> 17 and True
True
```

This flexibility can be useful, but there are some subtleties to it that might be confusing. You might want to avoid it (unless you know what you are doing).

Conditional Execution

In order to write useful programs, we almost always need the ability to check conditions and change the behavior of the program accordingly. *Conditional statements* give us this ability. The simplest form is the `if` statement:

```
if x > 0:
    print 'x is positive'
```

The boolean expression after `if` is called the **condition**. If it is true, then the indented statement gets executed. If not, nothing happens.

`if` statements have the same structure as function definitions: a header followed by an indented body. Statements like this are called **compound statements**.

There is no limit on the number of statements that can appear in the body, but there has to be at least one. Occasionally, it is useful to have a body with no statements (usually as a place keeper for code you haven't written yet). In that case, you can use the `pass` statement, which does nothing.

```
if x < 0:
    pass            # need to handle negative values!
```

Alternative Execution

A second form of the `if` statement is **alternative execution**, in which there are two possibilities and the condition determines which one gets executed. The syntax looks like this:

```
if x%2 == 0:
    print 'x is even'
else:
    print 'x is odd'
```

If the remainder when x is divided by 2 is 0, then we know that x is even, and the program displays a message to that effect. If the condition is false, the second set of statements is executed. Since the condition must be true or false, exactly one of the alternatives will be executed. The alternatives are called **branches**, because they are branches in the flow of execution.

Chained Conditionals

Sometimes there are more than two possibilities and we need more than two branches. One way to express a computation like that is a **chained conditional**:

```
if x < y:
    print 'x is less than y'
elif x > y:
    print 'x is greater than y'
else:
    print 'x and y are equal'
```

`elif` is an abbreviation of "else if." Again, exactly one branch will be executed. There is no limit on the number of `elif` statements. If there is an `else` clause, it has to be at the end, but there doesn't have to be one.

```
if choice == 'a':
    draw_a()
elif choice == 'b':
    draw_b()
elif choice == 'c':
    draw_c()
```

Each condition is checked in order. If the first is false, the next is checked, and so on. If one of them is true, the corresponding branch executes, and the statement ends. Even if more than one condition is true, only the first true branch executes.

Nested Conditionals

One conditional can also be nested within another. We could have written the trichotomy example like this:

```
if x == y:
    print 'x and y are equal'
else:
    if x < y:
        print 'x is less than y'
    else:
        print 'x is greater than y'
```

The outer conditional contains two branches. The first branch contains a simple statement. The second branch contains another if statement, which has two branches of its own. Those two branches are both simple statements, although they could have been conditional statements as well.

Although the indentation of the statements makes the structure apparent, **nested conditionals** become difficult to read very quickly. In general, it is a good idea to avoid them when you can.

Logical operators often provide a way to simplify nested conditional statements. For example, we can rewrite the following code using a single conditional:

```
if 0 < x:
    if x < 10:
        print 'x is a positive single-digit number.'
```

The print statement is executed only if we make it past both conditionals, so we can get the same effect with the and operator:

```
if 0 < x and x < 10:
    print 'x is a positive single-digit number.'
```

Recursion

It is legal for one function to call another; it is also legal for a function to call itself. It may not be obvious why that is a good thing, but it turns out to be one of the most magical things a program can do. For example, look at the following function:

```
def countdown(n):
    if n <= 0:
        print 'Blastoff!'
    else:
        print n
        countdown(n-1)
```

If n is 0 or negative, it outputs the word, "Blastoff!" Otherwise, it outputs n and then calls a function named countdown—itself—passing n-1 as an argument.

What happens if we call this function like this?

```
>>> countdown(3)
```

The execution of countdown begins with n=3, and since n is greater than 0, it outputs the value 3, and then calls itself...

> The execution of countdown begins with n=2, and since n is greater than 0, it outputs the value 2, and then calls itself...

> > The execution of countdown begins with n=1, and since n is greater than 0, it outputs the value 1, and then calls itself...

> > > The execution of countdown begins with n=0, and since n is not greater than 0, it outputs the word, "Blastoff!" and then returns.

> > The countdown that got n=1 returns.

> The countdown that got n=2 returns.

The countdown that got n=3 returns.

And then you're back in __main__. So, the total output looks like this:

```
3
2
1
Blastoff!
```

A function that calls itself is **recursive**; the process is called **recursion**.

As another example, we can write a function that prints a string n times.

```
def print_n(s, n):
    if n <= 0:
        return
    print s
    print_n(s, n-1)
```

If n <= 0 the return statement exits the function. The flow of execution immediately returns to the caller, and the remaining lines of the function are not executed.

The rest of the function is similar to countdown: if n is greater than 0, it displays s and then calls itself to display s *n-1* additional times. So the number of lines of output is 1 + (n - 1), which adds up to n.

For simple examples like this, it is probably easier to use a for loop. But we will see examples later that are hard to write with a for loop and easy to write with recursion, so it is good to start early.

Stack Diagrams for Recursive Functions

In "Stack Diagrams" (page 30), we used a stack diagram to represent the state of a program during a function call. The same kind of diagram can help interpret a recursive function.

Every time a function gets called, Python creates a new function frame, which contains the function's local variables and parameters. For a recursive function, there might be more than one frame on the stack at the same time.

Figure 5-1 shows a stack diagram for countdown called with n = 3.

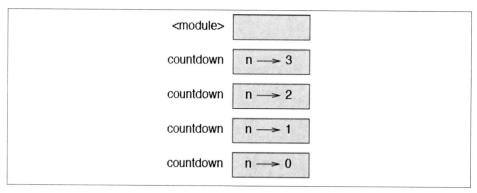

Figure 5-1. Stack diagram.

As usual, the top of the stack is the frame for __main__. It is empty because we did not create any variables in __main__ or pass any arguments to it.

The four countdown frames have different values for the parameter n. The bottom of the stack, where n=0, is called the **base case**. It does not make a recursive call, so there are no more frames.

Exercise 5-1.
Draw a stack diagram for print_n called with s = 'Hello' and n=2.

Exercise 5-2.
Write a function called do_n that takes a function object and a number, n, as arguments, and that calls the given function n times.

Infinite Recursion

If a recursion never reaches a base case, it goes on making recursive calls forever, and the program never terminates. This is known as **infinite recursion**, and it is generally not a good idea. Here is a minimal program with an infinite recursion:

```
def recurse():
    recurse()
```

In most programming environments, a program with infinite recursion does not really run forever. Python reports an error message when the maximum recursion depth is reached:

```
    File "<stdin>", line 2, in recurse
    File "<stdin>", line 2, in recurse
    File "<stdin>", line 2, in recurse
                    .
                    .
                    .
    File "<stdin>", line 2, in recurse
RuntimeError: Maximum recursion depth exceeded
```

This traceback is a little bigger than the one we saw in the previous chapter. When the error occurs, there are 1000 recurse frames on the stack!

Keyboard Input

The programs we have written so far are a bit rude in the sense that they accept no input from the user. They just do the same thing every time.

Python 2 provides a built-in function called raw_input that gets input from the keyboard. In Python 3, it is called input. When this function is called, the program stops and waits for the user to type something. When the user presses Return or Enter, the program resumes and raw_input returns what the user typed as a string.

```
>>> input = raw_input()
What are you waiting for?
>>> print input
What are you waiting for?
```

Before getting input from the user, it is a good idea to print a prompt telling the user what to input. raw_input can take a prompt as an argument:

```
>>> name = raw_input('What...is your name?\n')
What...is your name?
Arthur, King of the Britons!
>>> print name
Arthur, King of the Britons!
```

The sequence \n at the end of the prompt represents a **newline**, which is a special character that causes a line break. That's why the user's input appears below the prompt.

If you expect the user to type an integer, you can try to convert the return value to int:

```
>>> prompt = 'What...is the airspeed velocity of an unladen swallow?\n'
>>> speed = raw_input(prompt)
What...is the airspeed velocity of an unladen swallow?
17
>>> int(speed)
17
```

But if the user types something other than a string of digits, you get an error:

```
>>> speed = raw_input(prompt)
What...is the airspeed velocity of an unladen swallow?
What do you mean, an African or a European swallow?
>>> int(speed)
ValueError: invalid literal for int()
```

We will see how to handle this kind of error later.

Debugging

The traceback Python displays when an error occurs contains a lot of information, but it can be overwhelming, especially when there are many frames on the stack. The most useful parts are usually:

- What kind of error it was, and
- Where it occurred.

Syntax errors are usually easy to find, but there are a few gotchas. Whitespace errors can be tricky because spaces and tabs are invisible and we are used to ignoring them.

```
>>> x = 5
>>>  y = 6
```

```
  File "<stdin>", line 1
    y = 6
    ^
SyntaxError: invalid syntax
```

In this example, the problem is that the second line is indented by one space. But the error message points to y, which is misleading. In general, error messages indicate where the problem was discovered, but the actual error might be earlier in the code, sometimes on a previous line.

The same is true of runtime errors.

Suppose you are trying to compute a signal-to-noise ratio in decibels. The formula is $SNR_{db} = 10\log_{10}(P_{signal} / P_{noise})$. In Python, you might write something like this:

```
import math
signal_power = 9
noise_power = 10
ratio = signal_power / noise_power
decibels = 10 * math.log10(ratio)
print decibels
```

But when you run it in Python 2, you get an error message.

```
Traceback (most recent call last):
  File "snr.py", line 5, in ?
    decibels = 10 * math.log10(ratio)
OverflowError: math range error
```

The error message indicates line 5, but there is nothing wrong with that line. To find the real error, it might be useful to print the value of ratio, which turns out to be 0. The problem is in line 4, because dividing two integers does floor division. The solution is to represent signal power and noise power with floating-point values.

In general, error messages tell you where the problem was discovered, but that is often not where it was caused.

In Python 3, this example does not cause an error; the division operator performs floating-point division even with integer operands.

Glossary

Modulus operator:
> An operator, denoted with a percent sign (%), that works on integers and yields the remainder when one number is divided by another.

Boolean expression:
> An expression whose value is either True or False.

Relational operator:
> One of the operators that compares its operands: ==, !=, >, <, >=, and <=.

Logical operator:
> One of the operators that combines boolean expressions: and, or, and not.

Conditional statement:
> A statement that controls the flow of execution depending on some condition.

Condition:
> The boolean expression in a conditional statement that determines which branch is executed.

Compound statement:
> A statement that consists of a header and a body. The header ends with a colon (:). The body is indented relative to the header.

Branch:
> One of the alternative sequences of statements in a conditional statement.

Chained conditional:
> A conditional statement with a series of alternative branches.

Nested conditional:
> A conditional statement that appears in one of the branches of another conditional statement.

Recursion:
> The process of calling the function that is currently executing.

Base case:
> A conditional branch in a recursive function that does not make a recursive call.

Infinite recursion:
> A recursion that doesn't have a base case, or never reaches it. Eventually, an infinite recursion causes a runtime error.

Exercises

Exercise 5-3.

Fermat's Last Theorem says that there are no integers a, b, and c such that

$$a^n + b^n = c^n$$

for any values of n greater than 2.

1. Write a function named check_fermat that takes four parameters—a, b, c and n— and that checks to see if Fermat's theorem holds. If n is greater than 2 and it turns out to be true that

$$a^n + b^n = c^n$$

the program should print, "Holy smokes, Fermat was wrong!" Otherwise the program should print, "No, that doesn't work."

2. Write a function that prompts the user to input values for a, b, c and n, converts them to integers, and uses check_fermat to check whether they violate Fermat's theorem.

Exercise 5-4.
If you are given three sticks, you may or may not be able to arrange them in a triangle. For example, if one of the sticks is 12 inches long and the other two are one inch long, it is clear that you will not be able to get the short sticks to meet in the middle. For any three lengths, there is a simple test to see if it is possible to form a triangle:

> If any of the three lengths is greater than the sum of the other two, then you cannot form a triangle. Otherwise, you can. (If the sum of two lengths equals the third, they form what is called a "degenerate" triangle.)

1. Write a function named is_triangle that takes three integers as arguments, and that prints either "Yes" or "No," depending on whether you can or cannot form a triangle from sticks with the given lengths.

2. Write a function that prompts the user to input three stick lengths, converts them to integers, and uses is_triangle to check whether sticks with the given lengths can form a triangle.

The following exercises use TurtleWorld from Chapter 4:

Exercise 5-5.
Read the following function and see if you can figure out what it does. Then run it (see the examples in Chapter 4).

```
def draw(t, length, n):
    if n == 0:
        return
    angle = 50
    fd(t, length*n)
    lt(t, angle)
    draw(t, length, n-1)
    rt(t, 2*angle)
    draw(t, length, n-1)
    lt(t, angle)
    bk(t, length*n)
```

Exercise 5-6.
The Koch curve is a fractal that looks something like Figure 5-2. To draw a Koch curve with length *x*, all you have to do is:

1. Draw a Koch curve with length *x*/3.

2. Turn left 60 degrees.

3. Draw a Koch curve with length *x*/3.

4. Turn right 120 degrees.

5. Draw a Koch curve with length *x*/3.

6. Turn left 60 degrees.

7. Draw a Koch curve with length *x*/3.

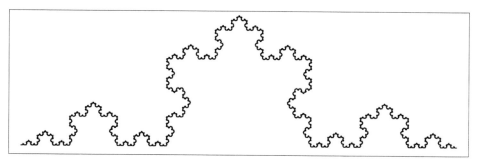

Figure 5-2. A Koch curve.

The exception is if *x* is less than 3: in that case, you can just draw a straight line with length *x*.

1. Write a function called koch that takes a turtle and a length as parameters, and that uses the turtle to draw a Koch curve with the given length.

2. Write a function called snowflake that draws three Koch curves to make the outline of a snowflake.

   Solution: *http://thinkpython.com/code/koch.py*.

3. The Koch curve can be generalized in several ways. See *http://en.wikipedia.org/wiki/Koch_snowflake* for examples and implement your favorite.

## CHAPTER 6

Fruitful Functions

Return Values

Some of the built-in functions we have used, such as the math functions, produce results. Calling the function generates a value, which we usually assign to a variable or use as part of an expression.

```
e = math.exp(1.0)
height = radius * math.sin(radians)
```

All of the functions we have written so far are void; they print something or move turtles around, but their return value is None.

In this chapter, we are (finally) going to write fruitful functions. The first example is area, which returns the area of a circle with the given radius:

```
def area(radius):
    temp = math.pi * radius**2
    return temp
```

We have seen the return statement before, but in a fruitful function the return statement includes an expression. This statement means: "Return immediately from this function and use the following expression as a return value." The expression can be arbitrarily complicated, so we could have written this function more concisely:

```
def area(radius):
    return math.pi * radius**2
```

On the other hand, **temporary variables** like temp often make debugging easier.

Sometimes it is useful to have multiple return statements, one in each branch of a conditional:

```
def absolute_value(x):
    if x < 0:
        return -x
    else:
        return x
```

Since these return statements are in an alternative conditional, only one will be executed.

As soon as a return statement executes, the function terminates without executing any subsequent statements. Code that appears after a return statement, or any other place the flow of execution can never reach, is called **dead code**.

In a fruitful function, it is a good idea to ensure that every possible path through the program hits a return statement. For example:

```
def absolute_value(x):
    if x < 0:
        return -x
    if x > 0:
        return x
```

This function is incorrect because if x happens to be 0, neither condition is true, and the function ends without hitting a return statement. If the flow of execution gets to the end of a function, the return value is None, which is not the absolute value of 0.

```
>>> print absolute_value(0)
None
```

By the way, Python provides a built-in function called abs that computes absolute values.

Exercise 6-1.

Write a compare function that returns 1 if x > y, 0 if x == y, and -1 if x < y.

Incremental Development

As you write larger functions, you might find yourself spending more time debugging.

To deal with increasingly complex programs, you might want to try a process called **incremental development**. The goal of incremental development is to avoid long debugging sessions by adding and testing only a small amount of code at a time.

As an example, suppose you want to find the distance between two points, given by the coordinates (x_1, y_1) and (x_2, y_2). By the Pythagorean theorem, the distance is:

$$distance = \sqrt{(x_2 - x_1)^2 + (y_2 - y_1)^2}$$

The first step is to consider what a distance function should look like in Python. In other words, what are the inputs (parameters) and what is the output (return value)?

In this case, the inputs are two points, which you can represent using four numbers. The return value is the distance, which is a floating-point value.

Already you can write an outline of the function:

```
def distance(x1, y1, x2, y2):
    return 0.0
```

Obviously, this version doesn't compute distances; it always returns zero. But it is syntactically correct, and it runs, which means that you can test it before you make it more complicated.

To test the new function, call it with sample arguments:

```
>>> distance(1, 2, 4, 6)
0.0
```

I chose these values so that the horizontal distance is 3 and the vertical distance is 4; that way, the result is 5 (the hypotenuse of a 3-4-5 triangle). When testing a function, it is useful to know the right answer.

At this point we have confirmed that the function is syntactically correct, and we can start adding code to the body. A reasonable next step is to find the differences $x_2 - x_1$ and $y_2 - y_1$. The next version stores those values in temporary variables and prints them.

```
def distance(x1, y1, x2, y2):
    dx = x2 - x1
    dy = y2 - y1
    print 'dx is', dx
    print 'dy is', dy
    return 0.0
```

If the function is working, it should display 'dx is 3' and 'dy is 4'. If so, we know that the function is getting the right arguments and performing the first computation correctly. If not, there are only a few lines to check.

Next we compute the sum of squares of dx and dy:

```
def distance(x1, y1, x2, y2):
    dx = x2 - x1
    dy = y2 - y1
    dsquared = dx**2 + dy**2
    print 'dsquared is: ', dsquared
    return 0.0
```

Again, you would run the program at this stage and check the output (which should be 25). Finally, you can use math.sqrt to compute and return the result:

```
def distance(x1, y1, x2, y2):
    dx = x2 - x1
    dy = y2 - y1
    dsquared = dx**2 + dy**2
    result = math.sqrt(dsquared)
    return result
```

If that works correctly, you are done. Otherwise, you might want to print the value of `result` before the return statement.

The final version of the function doesn't display anything when it runs; it only returns a value. The `print` statements we wrote are useful for debugging, but once you get the function working, you should remove them. Code like that is called **scaffolding** because it is helpful for building the program but is not part of the final product.

When you start out, you should add only a line or two of code at a time. As you gain more experience, you might find yourself writing and debugging bigger chunks. Either way, incremental development can save you a lot of debugging time.

The key aspects of the process are:

1. Start with a working program and make small incremental changes. At any point, if there is an error, you should have a good idea where it is.

2. Use temporary variables to hold intermediate values so you can display and check them.

3. Once the program is working, you might want to remove some of the scaffolding or consolidate multiple statements into compound expressions, but only if it does not make the program difficult to read.

Exercise 6-2.
Use incremental development to write a function called `hypotenuse` that returns the length of the hypotenuse of a right triangle given the lengths of the two legs as arguments. Record each stage of the development process as you go.

Composition

As you should expect by now, you can call one function from within another. This ability is called **composition**.

As an example, we'll write a function that takes two points, the center of the circle and a point on the perimeter, and computes the area of the circle.

Assume that the center point is stored in the variables `xc` and `yc`, and the perimeter point is in `xp` and `yp`. The first step is to find the radius of the circle, which is the distance between the two points. We just wrote a function, `distance`, that does that:

```
radius = distance(xc, yc, xp, yp)
```

The next step is to find the area of a circle with that radius; we just wrote that, too:

```
result = area(radius)
```

Encapsulating these steps in a function, we get:

```
def circle_area(xc, yc, xp, yp):
    radius = distance(xc, yc, xp, yp)
    result = area(radius)
    return result
```

The temporary variables `radius` and `result` are useful for development and debugging, but once the program is working, we can make it more concise by composing the function calls:

```
def circle_area(xc, yc, xp, yp):
    return area(distance(xc, yc, xp, yp))
```

Boolean Functions

Functions can return booleans, which is often convenient for hiding complicated tests inside functions.For example:

```
def is_divisible(x, y):
    if x % y == 0:
        return True
    else:
        return False
```

It is common to give boolean functions names that sound like yes/no questions; `is_divisible` returns either `True` or `False` to indicate whether x is divisible by y.

Here is an example:

```
>>>    is_divisible(6, 4)
False
>>>    is_divisible(6, 3)
True
```

The result of the == operator is a boolean, so we can write the function more concisely by returning it directly:

```
def is_divisible(x, y):
    return x % y == 0
```

Boolean functions are often used in conditional statements:

```
if is_divisible(x, y):
    print 'x is divisible by y'
```

It might be tempting to write something like:

```
if is_divisible(x, y) == True:
    print 'x is divisible by y'
```

But the extra comparison is unnecessary.

Exercise 6-3.
Write a function `is_between(x, y, z)` that returns `True` if $x \leq y \leq z$ or `False` otherwise.

More Recursion

We have only covered a small subset of Python, but you might be interested to know that this subset is a *complete* programming language, which means that anything that can be computed can be expressed in this language. Any program ever written could be rewritten using only the language features you have learned so far (actually, you would need a few commands to control devices like the keyboard, mouse, disks, etc., but that's all).

Proving that claim is a nontrivial exercise first accomplished by Alan Turing, one of the first computer scientists (some would argue that he was a mathematician, but a lot of early computer scientists started as mathematicians). Accordingly, it is known as the Turing Thesis. For a more complete (and accurate) discussion of the Turing Thesis, I recommend Michael Sipser's book *Introduction to the Theory of Computation*.

To give you an idea of what you can do with the tools you have learned so far, we'll evaluate a few recursively defined mathematical functions. A recursive definition is similar to a circular definition, in the sense that the definition contains a reference to the thing being defined. A truly circular definition is not very useful:

Vorpal:
    An adjective used to describe something that is vorpal.

If you saw that definition in the dictionary, you might be annoyed. On the other hand, if you looked up the definition of the factorial function, denoted with the symbol *!*, you might get something like this:

$$0! = 1$$
$$n! = n(n-1)!$$

This definition says that the factorial of 0 is 1, and the factorial of any other value, *n*, is *n* multiplied by the factorial of *n-1*.

So *3!* is 3 times *2!*, which is 2 times *1!*, which is 1 times *0!*. Putting it all together, *3!* equals 3 times 2 times 1 times 1, which is 6.

If you can write a recursive definition of something, you can usually write a Python program to evaluate it. The first step is to decide what the parameters should be. In this case it should be clear that `factorial` takes an integer:

```
def factorial(n):
```

If the argument happens to be 0, all we have to do is return 1:

```
def factorial(n):
    if n == 0:
        return 1
```

Otherwise, and this is the interesting part, we have to make a recursive call to find the factorial of *n-1* and then multiply it by *n*:

```
def factorial(n):
    if n == 0:
        return 1
    else:
        recurse = factorial(n-1)
        result = n * recurse
        return result
```

The flow of execution for this program is similar to the flow of countdown in "Recursion" (page 53). If we call factorial with the value 3:

Since 3 is not 0, we take the second branch and calculate the factorial of n-1...

> Since 2 is not 0, we take the second branch and calculate the factorial of n-1...

> > Since 1 is not 0, we take the second branch and calculate the factorial of n-1...

> > > Since 0 *is* 0, we take the first branch and return 1 without making any more recursive calls.

> > The return value (1) is multiplied by *n*, which is 1, and the result is returned.

> The return value (1) is multiplied by *n*, which is 2, and the result is returned.

The return value (2) is multiplied by *n*, which is 3, and the result, 6, becomes the return value of the function call that started the whole process.

Figure 6-1 shows what the stack diagram looks like for this sequence of function calls.

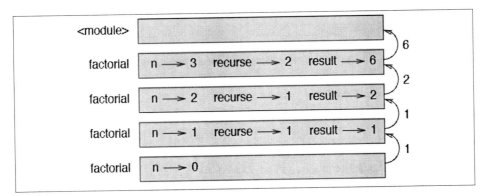

Figure 6-1. Stack diagram.

The return values are shown being passed back up the stack. In each frame, the return value is the value of result, which is the product of n and recurse.

In the last frame, the local variables recurse and result do not exist, because the branch that creates them does not execute.

Leap of Faith

Following the flow of execution is one way to read programs, but it can quickly become labyrinthine. An alternative is what I call the "leap of faith." When you come to a function call, instead of following the flow of execution, you *assume* that the function works correctly and returns the right result.

In fact, you are already practicing this leap of faith when you use built-in functions. When you call math.cos or math.exp, you don't examine the bodies of those functions. You just assume that they work because the people who wrote the built-in functions were good programmers.

The same is true when you call one of your own functions. For example, in "Boolean Functions" (page 65), we wrote a function called is_divisible that determines whether one number is divisible by another. Once we have convinced ourselves that this function is correct—by examining the code and testing—we can use the function without looking at the body again.

The same is true of recursive programs. When you get to the recursive call, instead of following the flow of execution, you should assume that the recursive call works (yields the correct result) and then ask yourself, "Assuming that I can find the factorial of *n-1*, can I compute the factorial of *n*?" In this case, it is clear that you can, by multiplying by *n*.

Of course, it's a bit strange to assume that the function works correctly when you haven't finished writing it, but that's why it's called a leap of faith!

One More Example

After factorial, the most common example of a recursively defined mathematical function is fibonacci, which has the following definition (see *http://en.wikipedia.org/wiki/Fibonacci_number*):

$$\text{fibonacci}(0) = 0$$
$$\text{fibonacci}(1) = 1$$
$$\text{fibonacci}(n) = \text{fibonacci}(n - 1) + \text{fibonacci}(n - 2)$$

Translated into Python, it looks like this:

```
def fibonacci (n):
    if n == 0:
        return 0
    elif  n == 1:
        return 1
    else:
        return fibonacci(n-1) + fibonacci(n-2)
```

If you try to follow the flow of execution here, even for fairly small values of *n*, your head explodes. But according to the leap of faith, if you assume that the two recursive calls work correctly, then it is clear that you get the right result by adding them together.

Checking Types

What happens if we call `factorial` and give it 1.5 as an argument?

```
>>> factorial(1.5)
RuntimeError: Maximum recursion depth exceeded
```

It looks like an infinite recursion. But how can that be? There is a base case—when n == 0. But if n is not an integer, we can *miss* the base case and recurse forever.

In the first recursive call, the value of n is 0.5. In the next, it is -0.5. From there, it gets smaller (more negative), but it will never be 0.

We have two choices. We can try to generalize the `factorial` function to work with floating-point numbers, or we can make `factorial` check the type of its argument. The first option is called the gamma function and it's a little beyond the scope of this book. So we'll go for the second.

We can use the built-in function `isinstance` to verify the type of the argument. While we're at it, we can also make sure the argument is positive:

```
def factorial (n):
    if not isinstance(n, int):
        print 'Factorial is only defined for integers.'
        return None
    elif n < 0:
        print 'Factorial is not defined for negative integers.'
        return None
    elif n == 0:
        return 1
    else:
        return n * factorial(n-1)
```

The first base case handles nonintegers; the second catches negative integers. In both cases, the program prints an error message and returns None to indicate that something went wrong:

```
>>> factorial('fred')
Factorial is only defined for integers.
None
>>> factorial(-2)
Factorial is not defined for negative integers.
None
```

If we get past both checks, then we know that *n* is positive or zero, so we can prove that the recursion terminates.

This program demonstrates a pattern sometimes called a **guardian**. The first two conditionals act as guardians, protecting the code that follows from values that might cause an error. The guardians make it possible to prove the correctness of the code.

In "Reverse Lookup" (page 125) we will see a more flexible alternative to printing an error message: raising an exception.

Debugging

Breaking a large program into smaller functions creates natural checkpoints for debugging.If a function is not working, there are three possibilities to consider:

- There is something wrong with the arguments the function is getting; a precondition is violated.
- There is something wrong with the function; a postcondition is violated.
- There is something wrong with the return value or the way it is being used.

To rule out the first possibility, you can add a print statement at the beginning of the function and display the values of the parameters (and maybe their types). Or you can write code that checks the preconditions explicitly.

If the parameters look good, add a print statement before each return statement that displays the return value. If possible, check the result by hand. Consider calling the function with values that make it easy to check the result, as in "Incremental Development" (page 62).

If the function seems to be working, look at the function call to make sure the return value is being used correctly (or used at all!).

Adding print statements at the beginning and end of a function can help make the flow of execution more visible. For example, here is a version of factorial with print statements:

```
def factorial(n):
    space = ' ' * (4 * n)
    print space, 'factorial', n
    if n == 0:
        print space, 'returning 1'
        return 1
    else:
        recurse = factorial(n-1)
        result = n * recurse
        print space, 'returning', result
        return result
```

space is a string of space characters that controls the indentation of the output. Here is the result of factorial(5):

```
                        factorial 5
                    factorial 4
                factorial 3
            factorial 2
        factorial 1
    factorial 0
    returning 1
        returning 1
            returning 2
                returning 6
                    returning 24
                        returning 120
```

If you are confused about the flow of execution, this kind of output can be helpful. It takes some time to develop effective scaffolding, but a little bit of scaffolding can save a lot of debugging.

Glossary

Temporary variable:
A variable used to store an intermediate value in a complex calculation.

Dead code:
Part of a program that can never be executed, often because it appears after a return statement.

None:
A special value returned by functions that have no return statement or a return statement without an argument.

Incremental development:
A program development plan intended to avoid debugging by adding and testing only a small amount of code at a time.

Scaffolding:
    Code that is used during program development but is not part of the final version.

Guardian:
    A programming pattern that uses a conditional statement to check for and handle circumstances that might cause an error.

Exercises

Exercise 6-4.

Draw a stack diagram for the following program. What does the program print? Solution: *http://thinkpython.com/code/stack_diagram.py.*

```
def b(z):
    prod = a(z, z)
    print z, prod
    return prod

def a(x, y):
    x = x + 1
    return x * y

def c(x, y, z):
    total = x + y + z
    square = b(total)**2
    return square

x = 1
y = x + 1
print c(x, y+3, x+y)
```

Exercise 6-5.

The Ackermann function, $A(m, n)$, is defined:

$$A(m, n) = \begin{cases} n + 1 & \text{if } m = 0 \\ A(m - 1, 1) & \text{if } m > 0 \text{ and } n = 0 \\ A(m - 1, A(m, n - 1)) & \text{if } m > 0 \text{ and } n > 0. \end{cases}$$

See *http://en.wikipedia.org/wiki/Ackermann_function.* Write a function named ack that evaluates Ackermann's function. Use your function to evaluate ack(3, 4), which should be 125. What happens for larger values of m and n? Solution: *http://thinkpython.com/code/ackermann.py.*

Exercise 6-6.
A palindrome is a word that is spelled the same backward and forward, like "noon" and "redivider." Recursively, a word is a palindrome if the first and last letters are the same and the middle is a palindrome.

The following are functions that take a string argument and return the first, last, and middle letters:

```python
def first(word):
    return word[0]

def last(word):
    return word[-1]

def middle(word):
    return word[1:-1]
```

We'll see how they work in Chapter 8.

1. Type these functions into a file named `palindrome.py` and test them out. What happens if you call `middle` with a string with two letters? One letter? What about the empty string, which is written `' '` and contains no letters?

2. Write a function called `is_palindrome` that takes a string argument and returns `True` if it is a palindrome and `False` otherwise. Remember that you can use the built-in function `len` to check the length of a string.

Solution: *http://thinkpython.com/code/palindrome_soln.py.*

Exercise 6-7.
A number, *a*, is a power of *b* if it is divisible by *b* and *a/b* is a power of *b*. Write a function called `is_power` that takes parameters a and b and returns `True` if a is a power of b. Note: you will have to think about the base case.

Exercise 6-8.
The greatest common divisor (GCD) of *a* and *b* is the largest number that divides both of them with no remainder.

One way to find the GCD of two numbers is Euclid's algorithm, which is based on the observation that if *r* is the remainder when *a* is divided by *b*, then $gcd(a, b) = gcd(b, r)$. As a base case, we can use $gcd(a, 0) = a$.

Write a function called `gcd` that takes parameters a and b and returns their greatest common divisor. If you need help, see *http://en.wikipedia.org/wiki/Euclidean_algorithm*.

Credit: This exercise is based on an example from Abelson and Sussman's *Structure and Interpretation of Computer Programs.*

## CHAPTER 7

Iteration

Multiple Assignment

As you may have discovered, it is legal to make more than one assignment to the same variable. A new assignment makes an existing variable refer to a new value (and stop referring to the old value).

```
bruce = 5
print bruce,
bruce = 7
print bruce
```

The output of this program is 5 7, because the first time bruce is printed, its value is 5, and the second time, its value is 7. The comma at the end of the first print statement suppresses the newline, which is why both outputs appear on the same line.

Figure 7-1 shows what **multiple assignment** looks like in a state diagram.

With multiple assignment it is especially important to distinguish between an assignment operation and a statement of equality. Because Python uses the equal sign (=) for assignment, it is tempting to interpret a statement like a = b as a statement of equality. It is not!

First, equality is a symmetric relation and assignment is not. For example, in mathematics, if $a=7$ then $7=a$. But in Python, the statement a = 7 is legal and 7 = a is not.

Furthermore, in mathematics, a statement of equality is either true or false, for all time. If $a=b$ now, then a will always equal b. In Python, an assignment statement can make two variables equal, but they don't have to stay that way:

```
a = 5
b = a    # a and b are now equal
a = 3    # a and b are no longer equal
```

The third line changes the value of a but does not change the value of b, so they are no longer equal.

Although multiple assignment is frequently helpful, you should use it with caution. If the values of variables change frequently, it can make the code difficult to read and debug.

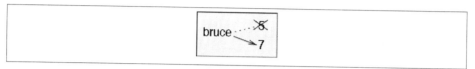

Figure 7-1. State diagram.

Updating Variables

One of the most common forms of multiple assignment is an **update**, where the new value of the variable depends on the old.

```
x = x+1
```

This means "get the current value of x, add one, and then update x with the new value."

If you try to update a variable that doesn't exist, you get an error, because Python evaluates the right side before it assigns a value to x:

```
>>> x = x+1
NameError: name 'x' is not defined
```

Before you can update a variable, you have to **initialize** it, usually with a simple assignment:

```
>>> x = 0
>>> x = x+1
```

Updating a variable by adding 1 is called an **increment**; subtracting 1 is called a **decrement**.

The while Statement

Computers are often used to automate repetitive tasks. Repeating identical or similar tasks without making errors is something that computers do well and people do poorly.

We have seen two programs, countdown and print_n, that use recursion to perform repetition, which is also called **iteration**. Because iteration is so common, Python provides several language features to make it easier. One is the for statement we saw in "Simple Repetition" (page 38). We'll get back to that later.

Another is the while statement. Here is a version of countdown that uses a while statement:

```
def countdown(n):
    while n > 0:
        print n
        n = n-1
    print 'Blastoff!'
```

You can almost read the while statement as if it were English. It means, "While n is greater than 0, display the value of n and then reduce the value of n by 1. When you get to 0, display the word Blastoff!"

More formally, here is the flow of execution for a while statement:

1. Evaluate the condition, yielding True or False.

2. If the condition is false, exit the while statement and continue execution at the next statement.

3. If the condition is true, execute the body and then go back to step 1.

This type of flow is called a **loop** because the third step loops back around to the top.

The body of the loop should change the value of one or more variables so that eventually the condition becomes false and the loop terminates. Otherwise the loop will repeat forever, which is called an **infinite loop**. An endless source of amusement for computer scientists is the observation that the directions on shampoo, "Lather, rinse, repeat," are an infinite loop.

In the case of countdown, we can prove that the loop terminates because we know that the value of n is finite, and we can see that the value of n gets smaller each time through the loop, so eventually we have to get to 0. In other cases, it is not so easy to tell:

```
def sequence(n):
    while n != 1:
        print n,
        if n%2 == 0:        # n is even
            n = n/2
        else:               # n is odd
            n = n*3+1
```

The condition for this loop is n != 1, so the loop will continue until n is 1, which makes the condition false.

Each time through the loop, the program outputs the value of n and then checks whether it is even or odd. If it is even, n is divided by 2. If it is odd, the value of n is replaced with n*3+1. For example, if the argument passed to sequence is 3, the resulting sequence is 3, 10, 5, 16, 8, 4, 2, 1.

Since n sometimes increases and sometimes decreases, there is no obvious proof that n will ever reach 1, or that the program terminates. For some particular values of n, we can prove termination. For example, if the starting value is a power of two, then the value of n will be even each time through the loop until it reaches 1. The previous example ends with such a sequence, starting with 16.

The hard question is whether we can prove that this program terminates for *all positive values* of n. So far, no one has been able to prove it *or* disprove it! (See *http://en.wikipedia.org/wiki/Collatz_conjecture.*)

Exercise 7-1.

Rewrite the function `print_n` from "Recursion" (page 53) using iteration instead of recursion.

break

Sometimes you don't know it's time to end a loop until you get half way through the body. In that case you can use the `break` statement to jump out of the loop.

For example, suppose you want to take input from the user until they type `done`. You could write:

```
while True:
    line = raw_input('> ')
    if line == 'done':
        break
    print line

print 'Done!'
```

The loop condition is `True`, which is always true, so the loop runs until it hits the break statement.

Each time through, it prompts the user with an angle bracket. If the user types `done`, the `break` statement exits the loop. Otherwise the program echoes whatever the user types and goes back to the top of the loop. Here's a sample run:

```
> not done
not done
> done
Done!
```

This way of writing `while` loops is common because you can check the condition anywhere in the loop (not just at the top) and you can express the stop condition affirmatively ("stop when this happens") rather than negatively ("keep going until that happens").

Square Roots

Loops are often used in programs that compute numerical results by starting with an approximate answer and iteratively improving it.

For example, one way of computing square roots is Newton's method. Suppose that you want to know the square root of a. If you start with almost any estimate, x, you can compute a better estimate with the following formula:

$$y = \frac{x + a / x}{2}$$

For example, if a is 4 and x is 3:

```
>>> a = 4.0
>>> x = 3.0
>>> y = (x + a/x) / 2
>>> print y
2.16666666667
```

Which is closer to the correct answer ($\sqrt{4} = 2$). If we repeat the process with the new estimate, it gets even closer:

```
>>> x = y
>>> y = (x + a/x) / 2
>>> print y
2.00641025641
```

After a few more updates, the estimate is almost exact:

```
>>> x = y
>>> y = (x + a/x) / 2
>>> print y
2.00001024003
>>> x = y
>>> y = (x + a/x) / 2
>>> print y
2.00000000003
```

In general we don't know ahead of time how many steps it takes to get to the right answer, but we know when we get there because the estimate stops changing:

```
>>> x = y
>>> y = (x + a/x) / 2
>>> print y
2.0
>>> x = y
>>> y = (x + a/x) / 2
>>> print y
2.0
```

When y == x, we can stop. Here is a loop that starts with an initial estimate, x, and improves it until it stops changing:

```
while True:
    print x
    y = (x + a/x) / 2
    if y == x:
        break
    x = y
```

For most values of a this works fine, but in general it is dangerous to test float equality. Floating-point values are only approximately right: most rational numbers, like *1/3*, and irrational numbers, like $\sqrt{2}$, can't be represented exactly with a float.

Rather than checking whether x and y are exactly equal, it is safer to use the built-in function abs to compute the absolute value, or magnitude, of the difference between them:

```
if abs(y-x) < epsilon:
    break
```

Where epsilon has a value like 0.0000001 that determines how close is close enough.

Exercise 7-2.
Encapsulate this loop in a function called square_root that takes a as a parameter, chooses a reasonable value of x, and returns an estimate of the square root of a.

Algorithms

Newton's method is an example of an **algorithm**: it is a mechanical process for solving a category of problems (in this case, computing square roots).

It is not easy to define an algorithm. It might help to start with something that is not an algorithm. When you learned to multiply single-digit numbers, you probably memorized the multiplication table. In effect, you memorized 100 specific solutions. That kind of knowledge is not algorithmic.

But if you were "lazy," you probably cheated by learning a few tricks. For example, to find the product of *n* and 9, you can write *n-1* as the first digit and *10-n* as the second digit. This trick is a general solution for multiplying any single-digit number by 9. That's an algorithm!

Similarly, the techniques you learned for addition with carrying, subtraction with borrowing, and long division are all algorithms. One of the characteristics of algorithms is that they do not require any intelligence to carry out. They are mechanical processes in which each step follows from the last according to a simple set of rules.

In my opinion, it is embarrassing that humans spend so much time in school learning to execute algorithms that, quite literally, require no intelligence.

On the other hand, the process of designing algorithms is interesting, intellectually challenging, and a central part of what we call programming.

Some of the things that people do naturally, without difficulty or conscious thought, are the hardest to express algorithmically. Understanding natural language is a good example. We all do it, but so far no one has been able to explain *how* we do it, at least not in the form of an algorithm.

Debugging

As you start writing bigger programs, you might find yourself spending more time debugging. More code means more chances to make an error and more place for bugs to hide.

One way to cut your debugging time is "debugging by bisection." For example, if there are 100 lines in your program and you check them one at a time, it would take 100 steps.

Instead, try to break the problem in half. Look at the middle of the program, or near it, for an intermediate value you can check. Add a `print` statement (or something else that has a verifiable effect) and run the program.

If the mid-point check is incorrect, there must be a problem in the first half of the program. If it is correct, the problem is in the second half.

Every time you perform a check like this, you halve the number of lines you have to search. After six steps (which is fewer than 100), you would be down to one or two lines of code, at least in theory.

In practice it is not always clear what the "middle of the program" is and not always possible to check it. It doesn't make sense to count lines and find the exact midpoint. Instead, think about places in the program where there might be errors and places where it is easy to put a check. Then choose a spot where you think the chances are about the same that the bug is before or after the check.

Glossary

Multiple assignment:
> Making more than one assignment to the same variable during the execution of a program.

Update:
> An assignment where the new value of the variable depends on the old.

Initialization:
> An assignment that gives an initial value to a variable that will be updated.

Increment:
    An update that increases the value of a variable (often by one).

Decrement:
    An update that decreases the value of a variable.

Iteration:
    Repeated execution of a set of statements using either a recursive function call or a loop.

Infinite loop:
    A loop in which the terminating condition is never satisfied.

Exercises

Exercise 7-3.

To test the square root algorithm in this chapter, you could compare it with `math.sqrt`. Write a function named `test_square_root` that prints a table like this:

```
1.0 1.0           1.0           0.0
2.0 1.41421356237 1.41421356237 2.22044604925e-16
3.0 1.73205080757 1.73205080757 0.0
4.0 2.0           2.0           0.0
5.0 2.2360679775  2.2360679775  0.0
6.0 2.44948974278 2.44948974278 0.0
7.0 2.64575131106 2.64575131106 0.0
8.0 2.82842712475 2.82842712475 4.4408920985e-16
9.0 3.0           3.0           0.0
```

The first column is a number, *a*; the second column is the square root of *a* computed with the function from "Square Roots" (page 79); the third column is the square root computed by `math.sqrt`; the fourth column is the absolute value of the difference between the two estimates.

Exercise 7-4.

The built-in function `eval` takes a string and evaluates it using the Python interpreter. For example:

```
>>> eval('1 + 2 * 3')
7
>>> import math
>>> eval('math.sqrt(5)')
2.2360679774997898
>>> eval('type(math.pi)')
<type 'float'>
```

Write a function called `eval_loop` that iteratively prompts the user, takes the resulting input and evaluates it using `eval`, and prints the result.

It should continue until the user enters 'done', and then return the value of the last expression it evaluated.

Exercise 7-5.
The mathematician Srinivasa Ramanujan found an infinite series that can be used to generate a numerical approximation of π:

$$\frac{1}{\pi} = \frac{2\sqrt{2}}{9801} \sum_{k=0}^{\infty} \frac{(4k)!(1103 + 26390k)}{(k!)^4 396^{4k}}$$

Write a function called `estimate_pi` that uses this formula to compute and return an estimate of π. It should use a `while` loop to compute terms of the summation until the last term is smaller than `1e-15` (which is Python notation for 10^{-15}). You can check the result by comparing it to `math.pi`.

Solution: *http://thinkpython.com/code/pi.py.*

## CHAPTER 8

Strings

A String Is a Sequence

A string is a **sequence** of characters. You can access the characters one at a time with the bracket operator:

```
>>> fruit = 'banana'
>>> letter = fruit[1]
```

The second statement selects character number 1 from `fruit` and assigns it to `letter`.

The expression in brackets is called an **index**. The index indicates which character in the sequence you want (hence the name).

But you might not get what you expect:

```
>>> print letter
a
```

For most people, the first letter of `'banana'` is b, not a. But for computer scientists, the index is an offset from the beginning of the string, and the offset of the first letter is zero.

```
>>> letter = fruit[0]
>>> print letter
b
```

So b is the 0th letter ("zero-eth") of `'banana'`, a is the 1th letter ("one-eth"), and n is the 2th ("two-eth") letter.

You can use any expression, including variables and operators, as an index, but the value of the index has to be an integer. Otherwise you get:

```
>>> letter = fruit[1.5]
TypeError: string indices must be integers
```

len

`len` is a built-in function that returns the number of characters in a string:

```
>>> fruit = 'banana'
>>> len(fruit)
6
```

To get the last letter of a string, you might be tempted to try something like this:

```
>>> length = len(fruit)
>>> last = fruit[length]
IndexError: string index out of range
```

The reason for the `IndexError` is that there is no letter in 'banana' with the index 6. Since we started counting at zero, the six letters are numbered 0 to 5. To get the last character, you have to subtract 1 from `length`:

```
>>> last = fruit[length-1]
>>> print last
a
```

Alternatively, you can use negative indices, which count backward from the end of the string. The expression `fruit[-1]` yields the last letter, `fruit[-2]` yields the second to last, and so on.

Traversal with a for Loop

A lot of computations involve processing a string one character at a time. Often they start at the beginning, select each character in turn, do something to it, and continue until the end. This pattern of processing is called a **traversal**. One way to write a traversal is with a `while` loop:

```
index = 0
while index < len(fruit):
    letter = fruit[index]
    print letter
    index = index + 1
```

This loop traverses the string and displays each letter on a line by itself. The loop condition is `index < len(fruit)`, so when `index` is equal to the length of the string, the condition is false, and the body of the loop is not executed. The last character accessed is the one with the index `len(fruit)-1`, which is the last character in the string.

Exercise 8-1.

Write a function that takes a string as an argument and displays the letters backward, one per line.

Another way to write a traversal is with a `for` loop:

```
for char in fruit:
    print char
```

Each time through the loop, the next character in the string is assigned to the variable char. The loop continues until no characters are left.

The following example shows how to use concatenation (string addition) and a for loop to generate an abecedarian series (that is, in alphabetical order). In Robert McCloskey's book *Make Way for Ducklings*, the names of the ducklings are Jack, Kack, Lack, Mack, Nack, Ouack, Pack, and Quack. This loop outputs these names in order:

```
prefixes = 'JKLMNOPQ'
suffix = 'ack'

for letter in prefixes:
    print letter + suffix
```

The output is:

```
Jack
Kack
Lack
Mack
Nack
Oack
Pack
Qack
```

Of course, that's not quite right because "Ouack" and "Quack" are misspelled.

Exercise 8-2.
Modify the program to fix this error.

String Slices

A segment of a string is called a **slice**. Selecting a slice is similar to selecting a character:

```
>>> s = 'Monty Python'
>>> print s[0:5]
Monty
>>> print s[6:12]
Python
```

The operator [n:m] returns the part of the string from the "n-eth" character to the "m-eth" character, including the first but excluding the last. This behavior is counter-intuitive, but it might help to imagine the indices pointing *between* the characters, as in Figure 8-1.

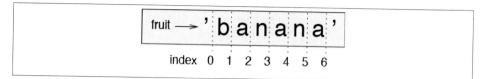

Figure 8-1. Slice indices.

If you omit the first index (before the colon), the slice starts at the beginning of the string. If you omit the second index, the slice goes to the end of the string:

```
>>> fruit = 'banana'
>>> fruit[:3]
'ban'
>>> fruit[3:]
'ana'
```

If the first index is greater than or equal to the second the result is an **empty string**, represented by two quotation marks:

```
>>> fruit = 'banana'
>>> fruit[3:3]
''
```

An empty string contains no characters and has length 0, but other than that, it is the same as any other string.

Exercise 8-3.
Given that `fruit` is a string, what does `fruit[:]` mean?

Strings Are Immutable

It is tempting to use the [ ] operator on the left side of an assignment, with the intention of changing a character in a string. For example:

```
>>> greeting = 'Hello, world!'
>>> greeting[0] = 'J'
TypeError: object does not support item assignment
```

The "object" in this case is the string and the "item" is the character you tried to assign. For now, an **object** is the same thing as a value, but we will refine that definition later. An **item** is one of the values in a sequence.

The reason for the error is that strings are **immutable**, which means you can't change an existing string. The best you can do is create a new string that is a variation on the original:

```
>>> greeting = 'Hello, world!'
>>> new_greeting = 'J' + greeting[1:]
>>> print new_greeting
Jello, world!
```

This example concatenates a new first letter onto a slice of `greeting`. It has no effect on the original string.

Searching

What does the following function do?

```
def find(word, letter):
    index = 0
    while index < len(word):
        if word[index] == letter:
            return index
        index = index + 1
    return -1
```

In a sense, `find` is the opposite of the `[]` operator. Instead of taking an index and extracting the corresponding character, it takes a character and finds the index where that character appears. If the character is not found, the function returns `-1`.

This is the first example we have seen of a `return` statement inside a loop. If `word[index] == letter`, the function breaks out of the loop and returns immediately.

If the character doesn't appear in the string, the program exits the loop normally and returns `-1`.

This pattern of computation—traversing a sequence and returning when we find what we are looking for—is called a **search**.

Exercise 8-4.
Modify `find` so that it has a third parameter, the index in `word` where it should start looking.

Looping and Counting

The following program counts the number of times the letter a appears in a string:

```
word = 'banana'
count = 0
for letter in word:
    if letter == 'a':
        count = count + 1
print count
```

This program demonstrates another pattern of computation called a **counter**. The variable `count` is initialized to 0 and then incremented each time an a is found. When the loop exits, `count` contains the result—the total number of a's.

Exercise 8-5.

Encapsulate this code in a function named `count`, and generalize it so that it accepts the string and the letter as arguments.

Exercise 8-6.

Rewrite this function so that instead of traversing the string, it uses the three-parameter version of `find` from the previous section.

String Methods

A **method** is similar to a function—it takes arguments and returns a value—but the syntax is different. For example, the method `upper` takes a string and returns a new string with all uppercase letters:

Instead of the function syntax `upper(word)`, it uses the method syntax `word.upper()`.

```
>>> word = 'banana'
>>> new_word = word.upper()
>>> print new_word
BANANA
```

This form of dot notation specifies the name of the method, `upper`, and the name of the string to apply the method to, `word`. The empty parentheses indicate that this method takes no argument.

A method call is called an **invocation**; in this case, we would say that we are invoking `upper` on the `word`.

As it turns out, there is a string method named `find` that is remarkably similar to the function we wrote:

```
>>> word = 'banana'
>>> index = word.find('a')
>>> print index
1
```

In this example, we invoke `find` on `word` and pass the letter we are looking for as a parameter.

Actually, the `find` method is more general than our function; it can find substrings, not just characters:

```
>>> word.find('na')
2
```

It can take as a second argument the index where it should start:

```
>>> word.find('na', 3)
4
```

And as a third argument the index where it should stop:

```
>>> name = 'bob'
>>> name.find('b', 1, 2)
-1
```

This search fails because b does not appear in the index range from 1 to 2 (not including 2).

Exercise 8-7.
There is a string method called count that is similar to the function in the previous exercise. Read the documentation of this method and write an invocation that counts the number of as in 'banana'.

Exercise 8-8.
Read the documentation of the string methods at *http://docs.python.org/lib/string-methods.html*. You might want to experiment with some of them to make sure you understand how they work. strip and replace are particularly useful.

The documentation uses a syntax that might be confusing. For example, the brackets in find(sub[, start[, end]]) indicate optional arguments. So sub is required, but start is optional, and if you include start, then end is optional.

The in Operator

The word in is a boolean operator that takes two strings and returns True if the first appears as a substring in the second:

```
>>> 'a' in 'banana'
True
>>> 'seed' in 'banana'
False
```

For example, the following function prints all the letters from word1 that also appear in word2:

```
def in_both(word1, word2):
    for letter in word1:
        if letter in word2:
            print letter
```

With well-chosen variable names, Python sometimes reads like English. You could read this loop, "for (each) letter in (the first) word, if (the) letter (appears) in (the second) word, print (the) letter."

Here's what you get if you compare apples and oranges:

```
>>> in_both('apples', 'oranges')
a
e
s
```

String Comparison

The relational operators work on strings. To see if two strings are equal:

```
if word == 'banana':
    print 'All right, bananas.'
```

Other relational operations are useful for putting words in alphabetical order:

```
if word < 'banana':
    print 'Your word,' + word + ', comes before banana.'
elif word > 'banana':
    print 'Your word,' + word + ', comes after banana.'
else:
    print 'All right, bananas.'
```

Python does not handle uppercase and lowercase letters the same way that people do. All the uppercase letters come before all the lowercase letters, so:

```
Your word, Pineapple, comes before banana.
```

A common way to address this problem is to convert strings to a standard format, such as all lowercase, before performing the comparison. Keep that in mind in case you have to defend yourself against a man armed with a Pineapple.

Debugging

When you use indices to traverse the values in a sequence, it is tricky to get the beginning and end of the traversal right. Here is a function that is supposed to compare two words and return True if one of the words is the reverse of the other, but it contains two errors:

```
def is_reverse(word1, word2):
    if len(word1) != len(word2):
        return False

    i = 0
    j = len(word2)

    while j > 0:
        if word1[i] != word2[j]:
            return False
        i = i+1
        j = j-1

    return True
```

The first `if` statement checks whether the words are the same length. If not, we can return `False` immediately and then, for the rest of the function, we can assume that the words are the same length. This is an example of the guardian pattern in "Checking Types" (page 69).

---

i and j are indices: i traverses word1 forward while j traverses word2 backward. If we find two letters that don't match, we can return False immediately. If we get through the whole loop and all the letters match, we return True.

If we test this function with the words "pots" and "stop", we expect the return value True, but we get an IndexError:

```
>>> is_reverse('pots', 'stop')
...
  File "reverse.py", line 15, in is_reverse
    if word1[i] != word2[j]:
IndexError: string index out of range
```

For debugging this kind of error, my first move is to print the values of the indices immediately before the line where the error appears.

```
    while j > 0:
        print i, j        # print here

        if word1[i] != word2[j]:
            return False
        i = i+1
        j = j-1
```

Now when I run the program again, I get more information:

```
>>> is_reverse('pots', 'stop')
0 4
...
IndexError: string index out of range
```

The first time through the loop, the value of j is 4, which is out of range for the string 'pots'. The index of the last character is 3, so the initial value for j should be len(word2)-1.

If I fix that error and run the program again, I get:

```
>>> is_reverse('pots', 'stop')
0 3
1 2
2 1
True
```

This time we get the right answer, but it looks like the loop only ran three times, which is suspicious. To get a better idea of what is happening, it is useful to draw a state diagram. During the first iteration, the frame for is_reverse is shows in Figure 8-2.

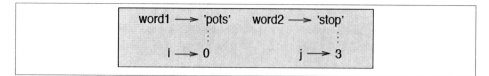

Figure 8-2. State diagram.

I took a little license by arranging the variables in the frame and adding dotted lines to show that the values of i and j indicate characters in word1 and word2.

Exercise 8-9.
Starting with this diagram, execute the program on paper, changing the values of i and j during each iteration. Find and fix the second error in this function.

Glossary

Object:
Something a variable can refer to. For now, you can use "object" and "value" interchangeably.

Sequence:
An ordered set; that is, a set of values where each value is identified by an integer index.

Item:
One of the values in a sequence.

Index:
An integer value used to select an item in a sequence, such as a character in a string.

Slice:
A part of a string specified by a range of indices.

Empty string:
A string with no characters and length 0, represented by two quotation marks.

Immutable:
The property of a sequence whose items cannot be assigned.

Traverse:
To iterate through the items in a sequence, performing a similar operation on each.

Search:
A pattern of traversal that stops when it finds what it is looking for.

Counter:
A variable used to count something, usually initialized to zero and then incremented.

Method:
  A function that is associated with an object and called using dot notation.

Invocation:
  A statement that calls a method.

Exercises

Exercise 8-10.
A string slice can take a third index that specifies the "step size;" that is, the number of spaces between successive characters. A step size of 2 means every other character; 3 means every third, etc.

```
>>> fruit = 'banana'
>>> fruit[0:5:2]
'bnn'
```

A step size of -1 goes through the word backwards, so the slice [::-1] generates a reversed string.

Use this idiom to write a one-line version of is_palindrome from Exercise 6-6.

Exercise 8-11.
The following functions are all *intended* to check whether a string contains any lowercase letters, but at least some of them are wrong. For each function, describe what the function actually does (assuming that the parameter is a string).

```
def any_lowercase1(s):
    for c in s:
        if c.islower():
            return True
        else:
            return False

def any_lowercase2(s):
    for c in s:
        if 'c'.islower():
            return 'True'
        else:
            return 'False'

def any_lowercase3(s):
    for c in s:
        flag = c.islower()
    return flag
```

```
def any_lowercase4(s):
    flag = False
    for c in s:
        flag = flag or c.islower()
    return flag

def any_lowercase5(s):
    for c in s:
        if not c.islower():
            return False
    return True
```

Exercise 8-12.

ROT13 is a weak form of encryption that involves "rotating" each letter in a word by 13 places. To rotate a letter means to shift it through the alphabet, wrapping around to the beginning if necessary, so 'A' shifted by 3 is 'D' and 'Z' shifted by 1 is 'A'.

Write a function called `rotate_word` that takes a string and an integer as parameters, and that returns a new string that contains the letters from the original string "rotated" by the given amount.

For example, "cheer" rotated by 7 is "jolly" and "melon" rotated by -10 is "cubed."

You might want to use the built-in functions `ord`, which converts a character to a numeric code, and `chr`, which converts numeric codes to characters.

Potentially offensive jokes on the Internet are sometimes encoded in ROT13. If you are not easily offended, find and decode some of them. Solution: *http://thinkpython.com/ code/rotate.py.*

## CHAPTER 9

Case Study: Word Play

Reading Word Lists

For the exercises in this chapter we need a list of English words. There are lots of word lists available on the Web, but the one most suitable for our purpose is one of the word lists collected and contributed to the public domain by Grady Ward as part of the Moby lexicon project (see *http://wikipedia.org/wiki/Moby_Project*). It is a list of 113,809 official crosswords; that is, words that are considered valid in crossword puzzles and other word games. In the Moby collection, the filename is 113809of.fic; you can download a copy, with the simpler name words.txt, from *http://thinkpython.com/code/words.txt*.

This file is in plain text, so you can open it with a text editor, but you can also read it from Python. The built-in function open takes the name of the file as a parameter and returns a **file object** you can use to read the file.

```
>>> fin = open('words.txt')
>>> print fin
<open file 'words.txt', mode 'r' at 0xb7f4b380>
```

fin is a common name for a file object used for input. Mode 'r' indicates that this file is open for reading (as opposed to 'w' for writing).

The file object provides several methods for reading, including readline, which reads characters from the file until it gets to a newline and returns the result as a string:

```
>>> fin.readline()
'aa\r\n'
```

The first word in this particular list, from *words.txt*, is "aa," which is a kind of lava. The sequence \r\n represents two whitespace characters, a carriage return and a newline, that separate this word from the next.

The file object keeps track of where it is in the file, so if you call `readline` again, you get the next word:

```
>>> fin.readline()
'aah\r\n'
```

The next word is "aah," which is a perfectly legitimate word, so stop looking at me like that. Or, if it's the whitespace that's bothering you, we can get rid of it with the string method `strip`:

```
>>> line = fin.readline()
>>> word = line.strip()
>>> print word
aahed
```

You can also use a file object as part of a `for` loop. This program reads `words.txt` and prints each word, one per line:

```
fin = open('words.txt')
for line in fin:
    word = line.strip()
    print word
```

Exercise 9-1.

Write a program that reads `words.txt` and prints only the words with more than 20 characters (not counting whitespace).

Exercises

There are solutions to these exercises in the next section. You should at least attempt each one before you read the solutions.

Exercise 9-2.

In 1939 Ernest Vincent Wright published a 50,000 word novel called *Gadsby* that does not contain the letter "e." Since "e" is the most common letter in English, that's not easy to do.

In fact, it is difficult to construct a solitary thought without using that most common symbol. It is slow going at first, but with caution and hours of training you can gradually gain facility.

All right, I'll stop now.

Write a function called `has_no_e` that returns `True` if the given word doesn't have the letter "e" in it.

Modify your program from the previous section to print only the words that have no "e" and compute the percentage of the words in the list have no "e."

Exercise 9-3.
Write a function named `avoids` that takes a word and a string of forbidden letters, and that returns `True` if the word doesn't use any of the forbidden letters.

Modify your program to prompt the user to enter a string of forbidden letters and then print the number of words that don't contain any of them. Can you find a combination of 5 forbidden letters that excludes the smallest number of words?

Exercise 9-4.
Write a function named `uses_only` that takes a word and a string of letters, and that returns `True` if the word contains only letters in the list. Can you make a sentence using only the letters `acefhlo`? Other than "Hoe alfalfa?"

Exercise 9-5.
Write a function named `uses_all` that takes a word and a string of required letters, and that returns `True` if the word uses all the required letters at least once. How many words are there that use all the vowels `aeiou`? How about `aeiouy`?

Exercise 9-6.
Write a function called `is_abecedarian` that returns `True` if the letters in a word appear in alphabetical order (double letters are ok). How many abecedarian words are there?

Search

All of the exercises in the previous section have something in common; they can be solved with the search pattern we saw in "Searching" (page 89). The simplest example is:

```
def has_no_e(word):
    for letter in word:
        if letter == 'e':
            return False
    return True
```

The `for` loop traverses the characters in `word`. If we find the letter "e", we can immediately return `False`; otherwise we have to go to the next letter. If we exit the loop normally, that means we didn't find an "e", so we return `True`.

`avoids` is a more general version of `has_no_e` but it has the same structure:

```
def avoids(word, forbidden):
    for letter in word:
        if letter in forbidden:
            return False
    return True
```

We can return `False` as soon as we find a forbidden letter; if we get to the end of the loop, we return `True`.

`uses_only` is similar except that the sense of the condition is reversed:

```
def uses_only(word, available):
    for letter in word:
        if letter not in available:
            return False
    return True
```

Instead of a list of forbidden letters, we have a list of available letters. If we find a letter in word that is not in `available`, we can return `False`.

`uses_all` is similar except that we reverse the role of the word and the string of letters:

```
def uses_all(word, required):
    for letter in required:
        if letter not in word:
            return False
    return True
```

Instead of traversing the letters in word, the loop traverses the required letters. If any of the required letters do not appear in the word, we can return `False`.

If you were really thinking like a computer scientist, you would have recognized that `uses_all` was an instance of a previously-solved problem, and you would have written:

```
def uses_all(word, required):
    return uses_only(required, word)
```

This is an example of a program development method called **problem recognition**, which means that you recognize the problem you are working on as an instance of a previously-solved problem, and apply a previously-developed solution.

Looping with Indices

I wrote the functions in the previous section with `for` loops because I only needed the characters in the strings; I didn't have to do anything with the indices.

For `is_abecedarian` we have to compare adjacent letters, which is a little tricky with a `for` loop:

```
def is_abecedarian(word):
    previous = word[0]
    for c in word:
        if c < previous:
            return False
        previous = c
    return True
```

An alternative is to use recursion:

```
def is_abecedarian(word):
    if len(word) <= 1:
```

```
        return True
    if word[0] > word[1]:
        return False
    return is_abecedarian(word[1:])
```

Another option is to use a while loop:

```
def is_abecedarian(word):
    i = 0
    while i < len(word)-1:
        if word[i+1] < word[i]:
            return False
        i = i+1
    return True
```

The loop starts at i=0 and ends when i=len(word)-1. Each time through the loop, it compares the *i*th character (which you can think of as the current character) to the *i* +*1*th character (which you can think of as the next).

If the next character is less than (alphabetically before) the current one, then we have discovered a break in the abecedarian trend, and we return False.

If we get to the end of the loop without finding a fault, then the word passes the test. To convince yourself that the loop ends correctly, consider an example like 'flossy'. The length of the word is 6, so the last time the loop runs is when i is 4, which is the index of the second-to-last character. On the last iteration, it compares the second-to-last character to the last, which is what we want.

Here is a version of is_palindrome (see Exercise 6-6) that uses two indices; one starts at the beginning and goes up; the other starts at the end and goes down.

```
def is_palindrome(word):
    i = 0
    j = len(word)-1

    while i<j:
        if word[i] != word[j]:
            return False
        i = i+1
        j = j-1

    return True
```

Or, if you noticed that this is an instance of a previously-solved problem, you might have written:

```
def is_palindrome(word):
    return is_reverse(word, word)
```

Assuming you did Exercise 8-9.

Debugging

Testing programs is hard. The functions in this chapter are relatively easy to test because you can check the results by hand. Even so, it is somewhere between difficult and impossible to choose a set of words that test for all possible errors.

Taking has_no_e as an example, there are two obvious cases to check: words that have an 'e' should return False; words that don't should return True. You should have no trouble coming up with one of each.

Within each case, there are some less obvious subcases. Among the words that have an "e," you should test words with an "e" at the beginning, the end, and somewhere in the middle. You should test long words, short words, and very short words, like the empty string. The empty string is an example of a **special case**, which is one of the non-obvious cases where errors often lurk.

In addition to the test cases you generate, you can also test your program with a word list like words.txt. By scanning the output, you might be able to catch errors, but be careful: you might catch one kind of error (words that should not be included, but are) and not another (words that should be included, but aren't).

In general, testing can help you find bugs, but it is not easy to generate a good set of test cases, and even if you do, you can't be sure your program is correct.

According to a legendary computer scientist:

> Program testing can be used to show the presence of bugs, but never to show their absence!
>
> —Edsger W. Dijkstra

Glossary

file object:
A value that represents an open file.

problem recognition:
A way of solving a problem by expressing it as an instance of a previously-solved problem.

special case:
A test case that is atypical or non-obvious (and less likely to be handled correctly).

Exercises

Exercise 9-7.
This question is based on a Puzzler that was broadcast on the radio program *Car Talk* (*http://www.cartalk.com/content/puzzler/transcripts/200725*):

> Give me a word with three consecutive double letters. I'll give you a couple of words that almost qualify, but don't. For example, the word committee, c-o-m-m-i-t-t-e-e. It would be great except for the 'i' that sneaks in there. Or Mississippi: M-i-s-s-i-s-s-i-p-p-i. If you could take out those i's it would work. But there is a word that has three consecutive pairs of letters and to the best of my knowledge this may be the only word. Of course there are probably 500 more but I can only think of one. What is the word?

Write a program to find it. Solution: *http://thinkpython.com/code/cartalk1.py*.

Exercise 9-8.
Here's another *Car Talk* Puzzler (*http://www.cartalk.com/content/puzzler/transcripts/200803*):

> "I was driving on the highway the other day and I happened to notice my odometer. Like most odometers, it shows six digits, in whole miles only. So, if my car had 300,000 miles, for example, I'd see 3-0-0-0-0-0.

> "Now, what I saw that day was very interesting. I noticed that the last 4 digits were palindromic; that is, they read the same forward as backward. For example, 5-4-4-5 is a palindrome, so my odometer could have read 3-1-5-4-4-5.

> "One mile later, the last 5 numbers were palindromic. For example, it could have read 3-6-5-4-5-6. One mile after that, the middle 4 out of 6 numbers were palindromic. And you ready for this? One mile later, all 6 were palindromic!

> "The question is, what was on the odometer when I first looked?"

Write a Python program that tests all the six-digit numbers and prints any numbers that satisfy these requirements. Solution: *http://thinkpython.com/code/cartalk2.py*.

Exercise 9-9.
Here's another *Car Talk* Puzzler you can solve with a search (*http://www.cartalk.com/content/puzzler/transcripts/200813*):

> "Recently I had a visit with my mom and we realized that the two digits that make up my age when reversed resulted in her age. For example, if she's 73, I'm 37. We wondered how often this has happened over the years but we got sidetracked with other topics and we never came up with an answer.

"When I got home I figured out that the digits of our ages have been reversible six times so far. I also figured out that if we're lucky it would happen again in a few years, and if we're really lucky it would happen one more time after that. In other words, it would have happened 8 times over all. So the question is, how old am I now?"

Write a Python program that searches for solutions to this Puzzler. Hint: you might find the string method zfill useful.

Solution: *http://thinkpython.com/code/cartalk3.py*.

## CHAPTER 10

Lists

A List Is a Sequence

Like a string, a **list** is a sequence of values. In a string, the values are characters; in a list, they can be any type. The values in a list are called **elements** or sometimes **items**.

There are several ways to create a new list; the simplest is to enclose the elements in square brackets ([and]):

```
[10, 20, 30, 40]
['crunchy frog', 'ram bladder', 'lark vomit']
```

The first example is a list of four integers. The second is a list of three strings. The elements of a list don't have to be the same type. The following list contains a string, a float, an integer, and (lo!) another list:

```
['spam', 2.0, 5, [10, 20]]
```

A list within another list is **nested**.

A list that contains no elements is called an empty list; you can create one with empty brackets, [].

As you might expect, you can assign list values to variables:

```
>>> cheeses = ['Cheddar', 'Edam', 'Gouda']
>>> numbers = [17, 123]
>>> empty = []
>>> print cheeses, numbers, empty
['Cheddar', 'Edam', 'Gouda'] [17, 123] []
```

Lists Are Mutable

The syntax for accessing the elements of a list is the same as for accessing the characters of a string—the bracket operator. The expression inside the brackets specifies the index. Remember that the indices start at 0:

```
>>> print cheeses[0]
Cheddar
```

Unlike strings, lists are mutable. When the bracket operator appears on the left side of an assignment, it identifies the element of the list that will be assigned.

```
>>> numbers = [17, 123]
>>> numbers[1] = 5
>>> print numbers
[17, 5]
```

The one-eth element of numbers, which used to be 123, is now 5.

You can think of a list as a relationship between indices and elements. This relationship is called a **mapping**; each index "maps to" one of the elements. Figure 10-1 shows the state diagram for cheeses, numbers and empty:

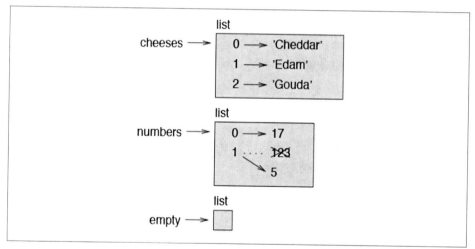

Figure 10-1. State diagram.

Lists are represented by boxes with the word "list" outside and the elements of the list inside. cheeses refers to a list with three elements indexed 0, 1 and 2. numbers contains two elements; the diagram shows that the value of the second element has been reassigned from 123 to 5. empty refers to a list with no elements.

List indices work the same way as string indices:

- Any integer expression can be used as an index.
- If you try to read or write an element that does not exist, you get an `IndexError`.
- If an index has a negative value, it counts backward from the end of the list.

The `in` operator also works on lists.

```
>>> cheeses = ['Cheddar', 'Edam', 'Gouda']
>>> 'Edam' in cheeses
True
>>> 'Brie' in cheeses
False
```

Traversing a List

The most common way to traverse the elements of a list is with a `for` loop. The syntax is the same as for strings:

```
for cheese in cheeses:
    print cheese
```

This works well if you only need to read the elements of the list. But if you want to write or update the elements, you need the indices. A common way to do that is to combine the functions `range` and `len`:

```
for i in range(len(numbers)):
    numbers[i] = numbers[i] * 2
```

This loop traverses the list and updates each element. `len` returns the number of elements in the list. `range` returns a list of indices from 0 to *n-1*, where *n* is the length of the list. Each time through the loop `i` gets the index of the next element. The assignment statement in the body uses `i` to read the old value of the element and to assign the new value.

A `for` loop over an empty list never executes the body:

```
for x in []:
    print 'This never happens.'
```

Although a list can contain another list, the nested list still counts as a single element. The length of this list is four:

```
['spam', 1, ['Brie', 'Roquefort', 'Pol le Veq'], [1, 2, 3]]
```

List Operations

The + operator concatenates lists:

```
>>> a = [1, 2, 3]
>>> b = [4, 5, 6]
>>> c = a + b
>>> print c
[1, 2, 3, 4, 5, 6]
```

Similarly, the * operator repeats a list a given number of times:

```
>>> [0] * 4
[0, 0, 0, 0]
>>> [1, 2, 3] * 3
[1, 2, 3, 1, 2, 3, 1, 2, 3]
```

The first example repeats [0] four times. The second example repeats the list [1, 2, 3] three times.

List Slices

The slice operator also works on lists:

```
>>> t = ['a', 'b', 'c', 'd', 'e', 'f']
>>> t[1:3]
['b', 'c']
>>> t[:4]
['a', 'b', 'c', 'd']
>>> t[3:]
['d', 'e', 'f']
```

If you omit the first index, the slice starts at the beginning. If you omit the second, the slice goes to the end. So if you omit both, the slice is a copy of the whole list.

```
>>> t[:]
['a', 'b', 'c', 'd', 'e', 'f']
```

Since lists are mutable, it is often useful to make a copy before performing operations that fold, spindle or mutilate lists.

A slice operator on the left side of an assignment can update multiple elements:

```
>>> t = ['a', 'b', 'c', 'd', 'e', 'f']
>>> t[1:3] = ['x', 'y']
>>> print t
['a', 'x', 'y', 'd', 'e', 'f']
```

List Methods

Python provides methods that operate on lists. For example, append adds a new element to the end of a list:

```
>>> t = ['a', 'b', 'c']
>>> t.append('d')
>>> print t
['a', 'b', 'c', 'd']
```

extend takes a list as an argument and appends all of the elements:

```
>>> t1 = ['a', 'b', 'c']
>>> t2 = ['d', 'e']
>>> t1.extend(t2)
>>> print t1
['a', 'b', 'c', 'd', 'e']
```

This example leaves t2 unmodified.

sort arranges the elements of the list from low to high:

```
>>> t = ['d', 'c', 'e', 'b', 'a']
>>> t.sort()
>>> print t
['a', 'b', 'c', 'd', 'e']
```

List methods are all void; they modify the list and return None. If you accidentally write t = t.sort(), you will be disappointed with the result.

Map, Filter, and Reduce

To add up all the numbers in a list, you can use a loop like this:

```
def add_all(t):
    total = 0
    for x in t:
        total += x
    return total
```

total is initialized to 0. Each time through the loop, x gets one element from the list. The += operator provides a short way to update a variable. This **augmented assignment statement**:

```
total += x
```

is equivalent to:

```
total = total + x
```

As the loop executes, total accumulates the sum of the elements; a variable used this way is sometimes called an **accumulator**.

Adding up the elements of a list is such a common operation that Python provides it as a built-in function, sum:

```
>>> t = [1, 2, 3]
>>> sum(t)
6
```

An operation like this that combines a sequence of elements into a single value is sometimes called **reduce**.

Exercise 10-1.

Write a function called `nested_sum` that takes a nested list of integers and add up the elements from all of the nested lists.

Sometimes you want to traverse one list while building another. For example, the following function takes a list of strings and returns a new list that contains capitalized strings:

```
def capitalize_all(t):
    res = []
    for s in t:
        res.append(s.capitalize())
    return res
```

`res` is initialized with an empty list; each time through the loop, we append the next element. So `res` is another kind of accumulator.

An operation like `capitalize_all` is sometimes called a **map** because it "maps" a function (in this case the method `capitalize`) onto each of the elements in a sequence.

Exercise 10-2.

Use `capitalize_all` to write a function named `capitalize_nested` that takes a nested list of strings and returns a new nested list with all strings capitalized.

Another common operation is to select some of the elements from a list and return a sublist. For example, the following function takes a list of strings and returns a list that contains only the uppercase strings:

```
def only_upper(t):
    res = []
    for s in t:
        if s.isupper():
            res.append(s)
    return res
```

`isupper` is a string method that returns `True` if the string contains only upper case letters.

An operation like `only_upper` is called a **filter** because it selects some of the elements and filters out the others.

Most common list operations can be expressed as a combination of map, filter and reduce. Because these operations are so common, Python provides language features to support them, including the built-in function `map` and an operator called a "list comprehension."

Exercise 10-3.

Write a function that takes a list of numbers and returns the cumulative sum; that is, a new list where the *i*th element is the sum of the first *i+1* elements from the original list. For example, the cumulative sum of `[1, 2, 3]` is `[1, 3, 6]`.

Deleting Elements

There are several ways to delete elements from a list. If you know the index of the element you want, you can use pop:

```
>>> t = ['a', 'b', 'c']
>>> x = t.pop(1)
>>> print t
['a', 'c']
>>> print x
b
```

pop modifies the list and returns the element that was removed. If you don't provide an index, it deletes and returns the last element.

If you don't need the removed value, you can use the del operator:

```
>>> t = ['a', 'b', 'c']
>>> del t[1]
>>> print t
['a', 'c']
```

If you know the element you want to remove (but not the index), you can use remove:

```
>>> t = ['a', 'b', 'c']
>>> t.remove('b')
>>> print t
['a', 'c']
```

The return value from remove is None.

To remove more than one element, you can use del with a slice index:

```
>>> t = ['a', 'b', 'c', 'd', 'e', 'f']
>>> del t[1:5]
>>> print t
['a', 'f']
```

As usual, the slice selects all the elements up to, but not including, the second index.

Exercise 10-4.
Write a function called middle that takes a list and returns a new list that contains all but the first and last elements. So middle([1,2,3,4]) should return [2,3].

Exercise 10-5.
Write a function called chop that takes a list, modifies it by removing the first and last elements, and returns None.

Lists and Strings

A string is a sequence of characters and a list is a sequence of values, but a list of characters is not the same as a string. To convert from a string to a list of characters, you can use list:

```
>>> s = 'spam'
>>> t = list(s)
>>> print t
['s', 'p', 'a', 'm']
```

Because list is the name of a built-in function, you should avoid using it as a variable name. I also avoid l because it looks too much like 1. So that's why I use t.

The list function breaks a string into individual letters. If you want to break a string into words, you can use the split method:

```
>>> s = 'pining for the fjords'
>>> t = s.split()
>>> print t
['pining', 'for', 'the', 'fjords']
```

An optional argument called a **delimiter** specifies which characters to use as word boundaries. The following example uses a hyphen as a delimiter:

```
>>> s = 'spam-spam-spam'
>>> delimiter = '-'
>>> s.split(delimiter)
['spam', 'spam', 'spam']
```

join is the inverse of split. It takes a list of strings and concatenates the elements. join is a string method, so you have to invoke it on the delimiter and pass the list as a parameter:

```
>>> t = ['pining', 'for', 'the', 'fjords']
>>> delimiter = ' '
>>> delimiter.join(t)
'pining for the fjords'
```

In this case the delimiter is a space character, so join puts a space between words. To concatenate strings without spaces, you can use the empty string, '', as a delimiter.

Objects and Values

If we execute these assignment statements:

```
a = 'banana'
b = 'banana'
```

We know that a and b both refer to a string, but we don't know whether they refer to the *same* string. There are two possible states, shown in Figure 10-2.

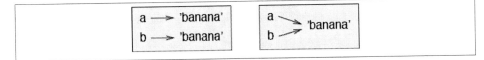

Figure 10-2. State diagram.

In one case, a and b refer to two different objects that have the same value. In the second case, they refer to the same object.

To check whether two variables refer to the same object, you can use the is operator.

```
>>> a = 'banana'
>>> b = 'banana'
>>> a is b
True
```

In this example, Python only created one string object, and both a and b refer to it.

But when you create two lists, you get two objects:

```
>>> a = [1, 2, 3]
>>> b = [1, 2, 3]
>>> a is b
False
```

So the state diagram looks like Figure 10-3.

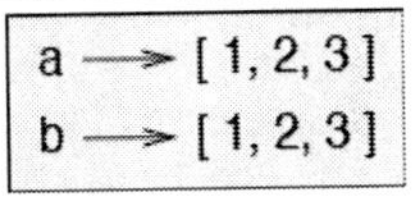

$$a \longrightarrow [\,1,2,3\,]$$
$$b \longrightarrow [\,1,2,3\,]$$

Figure 10-3. State diagram.

In this case we would say that the two lists are **equivalent**, because they have the same elements, but not **identical**, because they are not the same object. If two objects are identical, they are also equivalent, but if they are equivalent, they are not necessarily identical.

Until now, we have been using "object" and "value" interchangeably, but it is more precise to say that an object has a value. If you execute [1,2,3], you get a list object whose value is a sequence of integers. If another list has the same elements, we say it has the same value, but it is not the same object.

Aliasing

If a refers to an object and you assign b = a, then both variables refer to the same object:

```
>>> a = [1, 2, 3]
>>> b = a
>>> b is a
True
```

The state diagram looks like Figure 10-4.

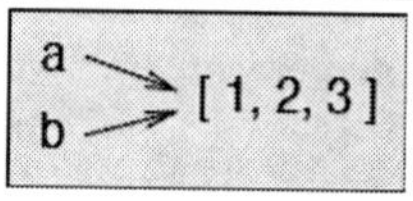

Figure 10-4. State diagram.

The association of a variable with an object is called a **reference**. In this example, there are two references to the same object.

An object with more than one reference has more than one name, so we say that the object is **aliased**.

If the aliased object is mutable, changes made with one alias affect the other:

```
>>> b[0] = 17
>>> print a
[17, 2, 3]
```

Although this behavior can be useful, it is error-prone. In general, it is safer to avoid aliasing when you are working with mutable objects.

For immutable objects like strings, aliasing is not as much of a problem. In this example:

```
a = 'banana'
b = 'banana'
```

It almost never makes a difference whether a and b refer to the same string or not.

List Arguments

When you pass a list to a function, the function gets a reference to the list. If the function modifies a list parameter, the caller sees the change. For example, delete_head removes the first element from a list:

```
def delete_head(t):
    del t[0]
```

Here's how it is used:

```
>>> letters = ['a', 'b', 'c']
>>> delete_head(letters)
>>> print letters
['b', 'c']
```

The parameter t and the variable letters are aliases for the same object. The stack diagram looks like Figure 10-5.

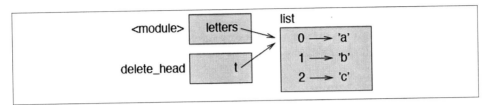

Figure 10-5. Stack diagram.

Since the list is shared by two frames, I drew it between them.

It is important to distinguish between operations that modify lists and operations that create new lists. For example, the append method modifies a list, but the + operator creates a new list:

```
>>> t1 = [1, 2]
>>> t2 = t1.append(3)
>>> print t1
[1, 2, 3]
>>> print t2
None

>>> t3 = t1 + [4]
>>> print t3
[1, 2, 3, 4]
```

This difference is important when you write functions that are supposed to modify lists. For example, this function *does not* delete the head of a list:

```
def bad_delete_head(t):
    t = t[1:]            # WRONG!
```

The slice operator creates a new list and the assignment makes t refer to it, but none of that has any effect on the list that was passed as an argument.

An alternative is to write a function that creates and returns a new list. For example, tail returns all but the first element of a list:

```
def tail(t):
    return t[1:]
```

This function leaves the original list unmodified. Here's how it is used:

```
>>> letters = ['a', 'b', 'c']
>>> rest = tail(letters)
>>> print rest
['b', 'c']
```

Debugging

Careless use of lists (and other mutable objects) can lead to long hours of debugging. Here are some common pitfalls and ways to avoid them:

1. Don't forget that most list methods modify the argument and return None. This is the opposite of the string methods, which return a new string and leave the original alone.

   If you are used to writing string code like this:

   ```
   word = word.strip()
   ```

   It is tempting to write list code like this:

   ```
   t = t.sort()          # WRONG!
   ```

   Because sort returns None, the next operation you perform with t is likely to fail.

   Before using list methods and operators, you should read the documentation carefully and then test them in interactive mode. The methods and operators that lists share with other sequences (like strings) are documented at *http://docs.python.org/lib/typesseq.html*. The methods and operators that only apply to mutable sequences are documented at *http://docs.python.org/lib/typesseq-mutable.html*.

2. Pick an idiom and stick with it.

   Part of the problem with lists is that there are too many ways to do things. For example, to remove an element from a list, you can use pop, remove, del, or even a slice assignment.

   To add an element, you can use the append method or the + operator. Assuming that t is a list and x is a list element, these are right:

   ```
   t.append(x)
   t = t + [x]
   ```

   And these are wrong:

   ```
   t.append([x])          # WRONG!
   t = t.append(x)        # WRONG!
   t + [x]                # WRONG!
   t = t + x              # WRONG!
   ```

   Try out each of these examples in interactive mode to make sure you understand what they do. Notice that only the last one causes a runtime error; the other three are legal, but they do the wrong thing.

3. Make copies to avoid aliasing.

   If you want to use a method like sort that modifies the argument, but you need to keep the original list as well, you can make a copy.

   ```
   orig = t[:]
   t.sort()
   ```

In this example you could also use the built-in function `sorted`, which returns a new, sorted list and leaves the original alone. But in that case you should avoid using `sorted` as a variable name!

Glossary

List:
> A sequence of values.

Element:
> One of the values in a list (or other sequence), also called items.

Index:
> An integer value that indicates an element in a list.

Nested list:
> A list that is an element of another list.

List traversal:
> The sequential accessing of each element in a list.

Mapping:
> A relationship in which each element of one set corresponds to an element of another set. For example, a list is a mapping from indices to elements.

Accumulator:
> A variable used in a loop to add up or accumulate a result.

Augmented assignment:
> A statement that updates the value of a variable using an operator like +=.

Reduce:
> A processing pattern that traverses a sequence and accumulates the elements into a single result.

Map:
> A processing pattern that traverses a sequence and performs an operation on each element.

Filter:
> A processing pattern that traverses a list and selects the elements that satisfy some criterion.

Object:
> Something a variable can refer to. An object has a type and a value.

Equivalent:
> Having the same value.

Identical:
    Being the same object (which implies equivalence).

Reference:
    The association between a variable and its value.

Aliasing:
    A circumstance where two or more variables refer to the same object.

Delimiter:
    A character or string used to indicate where a string should be split.

Exercises

Exercise 10-6.
Write a function called is_sorted that takes a list as a parameter and returns True if the list is sorted in ascending order and False otherwise. You can assume (as a precondition) that the elements of the list can be compared with the relational operators <, >, etc.

For example, is_sorted([1,2,2]) should return True and is_sorted(['b','a']) should return False.

Exercise 10-7.
Two words are anagrams if you can rearrange the letters from one to spell the other. Write a function called is_anagram that takes two strings and returns True if they are anagrams.

Exercise 10-8.
The (so-called) Birthday Paradox:

1. Write a function called has_duplicates that takes a list and returns True if there is any element that appears more than once. It should not modify the original list.

2. If there are 23 students in your class, what are the chances that two of you have the same birthday? You can estimate this probability by generating random samples of 23 birthdays and checking for matches. Hint: you can generate random birthdays with the randint function in the random module.

You can read about this problem at *http://en.wikipedia.org/wiki/Birthday_paradox*, and you can download my solution from *http://thinkpython.com/code/birthday.py*.

Exercise 10-9.
Write a function called remove_duplicates that takes a list and returns a new list with only the unique elements from the original. Hint: they don't have to be in the same order.

Exercise 10-10.
Write a function that reads the file words.txt and builds a list with one element per word. Write two versions of this function, one using the append method and the other using the idiom t = t + [x]. Which one takes longer to run? Why?

Hint: use the time module to measure elapsed time. Solution: *http://thinkpython.com/code/wordlist.py*.

Exercise 10-11.
To check whether a word is in the word list, you could use the in operator, but it would be slow because it searches through the words in order.

Because the words are in alphabetical order, we can speed things up with a bisection search (also known as binary search), which is similar to what you do when you look a word up in the dictionary. You start in the middle and check to see whether the word you are looking for comes before the word in the middle of the list. If so, then you search the first half of the list the same way. Otherwise you search the second half.

Either way, you cut the remaining search space in half. If the word list has 113,809 words, it will take about 17 steps to find the word or conclude that it's not there.

Write a function called bisect that takes a sorted list and a target value and returns the index of the value in the list, if it's there, or None if it's not.

Or you could read the documentation of the bisect module and use that! Solution: *http://thinkpython.com/code/inlist.py*.

Exercise 10-12.
Two words are a "reverse pair" if each is the reverse of the other. Write a program that finds all the reverse pairs in the word list. Solution: *http://thinkpython.com/code/reverse_pair.py*.

Exercise 10-13.
Two words "interlock" if taking alternating letters from each forms a new word. For example, "shoe" and "cold" interlock to form "schooled." Solution: *http://thinkpython.com/code/interlock.py*. Credit: This exercise is inspired by an example at *http://puzzlers.org*.

1. Write a program that finds all pairs of words that interlock. Hint: don't enumerate all pairs!

2. Can you find any words that are three-way interlocked; that is, every third letter forms a word, starting from the first, second or third?

## CHAPTER 11

Dictionaries

A **dictionary** is like a list, but more general. In a list, the indices have to be integers; in a dictionary they can be (almost) any type.

You can think of a dictionary as a mapping between a set of indices (which are called **keys**) and a set of values. Each key maps to a value. The association of a key and a value is called a **key-value pair** or sometimes an **item**.

As an example, we'll build a dictionary that maps from English to Spanish words, so the keys and the values are all strings.

The function `dict` creates a new dictionary with no items. Because `dict` is the name of a built-in function, you should avoid using it as a variable name.

```
>>> eng2sp = dict()
>>> print eng2sp
{}
```

The squiggly-brackets, {}, represent an empty dictionary. To add items to the dictionary, you can use square brackets:

```
>>> eng2sp['one'] = 'uno'
```

This line creates an item that maps from the key **'one'** to the value `'uno'`. If we print the dictionary again, we see a key-value pair with a colon between the key and value:

```
>>> print eng2sp
{'one': 'uno'}
```

This output format is also an input format. For example, you can create a new dictionary with three items:

```
>>> eng2sp = {'one': 'uno', 'two': 'dos', 'three': 'tres'}
```

But if you print eng2sp, you might be surprised:

```
>>> print eng2sp
{'one': 'uno', 'three': 'tres', 'two': 'dos'}
```

The order of the key-value pairs is not the same. In fact, if you type the same example on your computer, you might get a different result. In general, the order of items in a dictionary is unpredictable.

But that's not a problem because the elements of a dictionary are never indexed with integer indices. Instead, you use the keys to look up the corresponding values:

```
>>> print eng2sp['two']
'dos'
```

The key 'two' always maps to the value 'dos' so the order of the items doesn't matter.

If the key isn't in the dictionary, you get an exception:

```
>>> print eng2sp['four']
KeyError: 'four'
```

The len function works on dictionaries; it returns the number of key-value pairs:

```
>>> len(eng2sp)
3
```

The in operator works on dictionaries; it tells you whether something appears as a *key* in the dictionary (appearing as a value is not good enough).

```
>>> 'one' in eng2sp
True
>>> 'uno' in eng2sp
False
```

To see whether something appears as a value in a dictionary, you can use the method values, which returns the values as a list, and then use the in operator:

```
>>> vals = eng2sp.values()
>>> 'uno' in vals
True
```

The in operator uses different algorithms for lists and dictionaries. For lists, it uses a search algorithm, as in "Searching" (page 89). As the list gets longer, the search time gets longer in direct proportion. For dictionaries, Python uses an algorithm called a **hasht-able** that has a remarkable property: the in operator takes about the same amount of time no matter how many items there are in a dictionary. I won't explain how that's possible, but you can read more about it at *http://en.wikipedia.org/wiki/Hash_table*.

Exercise 11-1.
Write a function that reads the words in words.txt and stores them as keys in a dictionary. It doesn't matter what the values are. Then you can use the in operator as a fast way to check whether a string is in the dictionary.

If you did Exercise 10-11, you can compare the speed of this implementation with the list in operator and the bisection search.

Dictionary as a Set of Counters

Suppose you are given a string and you want to count how many times each letter appears. There are several ways you could do it:

1. You could create 26 variables, one for each letter of the alphabet. Then you could traverse the string and, for each character, increment the corresponding counter, probably using a chained conditional.

2. You could create a list with 26 elements. Then you could convert each character to a number (using the built-in function ord), use the number as an index into the list, and increment the appropriate counter.

3. You could create a dictionary with characters as keys and counters as the corresponding values. The first time you see a character, you would add an item to the dictionary. After that you would increment the value of an existing item.

Each of these options performs the same computation, but each of them implements that computation in a different way.

An **implementation** is a way of performing a computation; some implementations are better than others. For example, an advantage of the dictionary implementation is that we don't have to know ahead of time which letters appear in the string and we only have to make room for the letters that do appear.

Here is what the code might look like:

```
def histogram(s):
    d = dict()
    for c in s:
        if c not in d:
            d[c] = 1
        else:
            d[c] += 1
    return d
```

The name of the function is **histogram**, which is a statistical term for a set of counters (or frequencies).

The first line of the function creates an empty dictionary. The for loop traverses the string. Each time through the loop, if the character c is not in the dictionary, we create a new item with key c and the initial value 1 (since we have seen this letter once). If c is already in the dictionary we increment d[c].

Here's how it works:

```
>>> h = histogram('brontosaurus')
>>> print h
{'a': 1, 'b': 1, 'o': 2, 'n': 1, 's': 2, 'r': 2, 'u': 2, 't': 1}
```

The histogram indicates that the letters 'a' and 'b' appear once; 'o' appears twice, and so on.

Exercise 11-2.

Dictionaries have a method called get that takes a key and a default value. If the key appears in the dictionary, get returns the corresponding value; otherwise it returns the default value. For example:

```
>>> h = histogram('a')
>>> print h
{'a': 1}
>>> h.get('a', 0)
1
>>> h.get('b', 0)
0
```

Use get to write histogram more concisely. You should be able to eliminate the if statement.

Looping and Dictionaries

If you use a dictionary in a for statement, it traverses the keys of the dictionary. For example, print_hist prints each key and the corresponding value:

```
def print_hist(h):
    for c in h:
        print c, h[c]
```

Here's what the output looks like:

```
>>> h = histogram('parrot')
>>> print_hist(h)
a 1
p 1
r 2
t 1
o 1
```

Again, the keys are in no particular order.

Exercise 11-3.

Dictionaries have a method called keys that returns the keys of the dictionary, in no particular order, as a list.

Modify print_hist to print the keys and their values in alphabetical order.

Reverse Lookup

Given a dictionary d and a key k, it is easy to find the corresponding value v = d[k]. This operation is called a **lookup**.

But what if you have v and you want to find k? You have two problems: first, there might be more than one key that maps to the value v. Depending on the application, you might be able to pick one, or you might have to make a list that contains all of them. Second, there is no simple syntax to do a **reverse lookup**; you have to search.

Here is a function that takes a value and returns the first key that maps to that value:

```
def reverse_lookup(d, v):
    for k in d:
        if d[k] == v:
            return k
    raise ValueError
```

This function is yet another example of the search pattern, but it uses a feature we haven't seen before, raise. The raise statement causes an exception; in this case it causes a ValueError, which generally indicates that there is something wrong with the value of a parameter.

If we get to the end of the loop, that means v doesn't appear in the dictionary as a value, so we raise an exception.

Here is an example of a successful reverse lookup:

```
>>> h = histogram('parrot')
>>> k = reverse_lookup(h, 2)
>>> print k
r
```

And an unsuccessful one:

```
>>> k = reverse_lookup(h, 3)
Traceback (most recent call last):
  File "<stdin>", line 1, in ?
  File "<stdin>", line 5, in reverse_lookup
ValueError
```

The result when you raise an exception is the same as when Python raises one: it prints a traceback and an error message.

The raise statement takes a detailed error message as an optional argument. For example:

```
>>> raise ValueError, 'value does not appear in the dictionary'
Traceback (most recent call last):
  File "<stdin>", line 1, in ?
ValueError: value does not appear in the dictionary
```

A reverse lookup is much slower than a forward lookup; if you have to do it often, or if the dictionary gets big, the performance of your program will suffer.

Exercise 11-4.
Modify reverse_lookup so that it builds and returns a list of *all* keys that map to v, or an empty list if there are none.

Dictionaries and Lists

Lists can appear as values in a dictionary. For example, if you were given a dictionary that maps from letters to frequencies, you might want to invert it; that is, create a dictionary that maps from frequencies to letters. Since there might be several letters with the same frequency, each value in the inverted dictionary should be a list of letters.

Here is a function that inverts a dictionary:

```
def invert_dict(d):
    inverse = dict()
    for key in d:
        val = d[key]
        if val not in inverse:
            inverse[val] = [key]
        else:
            inverse[val].append(key)
    return inverse
```

Each time through the loop, key gets a key from d and val gets the corresponding value. If val is not in inverse, that means we haven't seen it before, so we create a new item and initialize it with a **singleton** (a list that contains a single element). Otherwise we have seen this value before, so we append the corresponding key to the list.

Here is an example:

```
>>> hist = histogram('parrot')
>>> print hist
{'a': 1, 'p': 1, 'r': 2, 't': 1, 'o': 1}
>>> inverse = invert_dict(hist)
>>> print inverse
{1: ['a', 'p', 't', 'o'], 2: ['r']}
```

Figure 11-1 is a state diagram showing hist and inverse. A dictionary is represented as a box with the type dict above it and the key-value pairs inside. If the values are integers, floats or strings, I usually draw them inside the box, but I usually draw lists outside the box, just to keep the diagram simple.

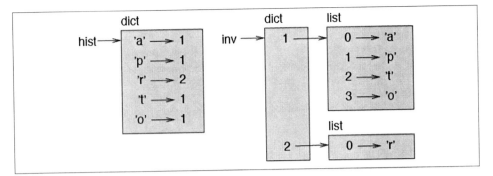

Figure 11-1. State diagram.

Lists can be values in a dictionary, as this example shows, but they cannot be keys. Here's what happens if you try:

```
>>> t = [1, 2, 3]
>>> d = dict()
>>> d[t] = 'oops'
Traceback (most recent call last):
  File "<stdin>", line 1, in ?
TypeError: list objects are unhashable
```

I mentioned earlier that a dictionary is implemented using a hashtable and that means that the keys have to be **hashable**.

A **hash** is a function that takes a value (of any kind) and returns an integer. Dictionaries use these integers, called hash values, to store and look up key-value pairs.

This system works fine if the keys are immutable. But if the keys are mutable, like lists, bad things happen. For example, when you create a key-value pair, Python hashes the key and stores it in the corresponding location. If you modify the key and then hash it again, it would go to a different location. In that case you might have two entries for the same key, or you might not be able to find a key. Either way, the dictionary wouldn't work correctly.

That's why the keys have to be hashable, and why mutable types like lists aren't. The simplest way to get around this limitation is to use tuples, which we will see in the next chapter.

Since dictionaries are mutable, they can't be used as keys, but they *can* be used as values.

Exercise 11-5.
Read the documentation of the dictionary method `setdefault` and use it to write a more concise version of `invert_dict`. Solution: *http://thinkpython.com/code/invert_dict.py*.

Memos

If you played with the fibonacci function from "One More Example" (page 68), you might have noticed that the bigger the argument you provide, the longer the function takes to run. Furthermore, the run time increases very quickly.

To understand why, consider Figure 11-2, which shows the **call graph** for fibonacci with n=4:

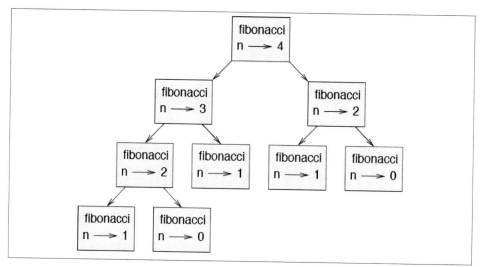

Figure 11-2. Call graph.

A call graph shows a set of function frames, with lines connecting each frame to the frames of the functions it calls. At the top of the graph, fibonacci with n=4 calls fibonacci with n=3 and n=2. In turn, fibonacci with n=3 calls fibonacci with n=2 and n=1. And so on.

Count how many times fibonacci(0) and fibonacci(1) are called. This is an inefficient solution to the problem, and it gets worse as the argument gets bigger.

One solution is to keep track of values that have already been computed by storing them in a dictionary. A previously computed value that is stored for later use is called a **memo**. Here is an implementation of fibonacci using memos:

```
known = {0:0, 1:1}

def fibonacci(n):
    if n in known:
        return known[n]

    res = fibonacci(n-1) + fibonacci(n-2)
    known[n] = res
    return res
```

known is a dictionary that keeps track of the Fibonacci numbers we already know. It starts with two items: 0 maps to 0 and 1 maps to 1.

Whenever fibonacci is called, it checks known. If the result is already there, it can return immediately. Otherwise it has to compute the new value, add it to the dictionary, and return it.

Exercise 11-6.
Run this version of fibonacci and the original with a range of parameters and compare their run times.

Exercise 11-7.
Memoize the Ackermann function from Exercise 6-5 and see if memoization makes it possible to evaluate the function with bigger arguments. Hint: no. Solution: *http://thinkpython.com/code/ackermann_memo.py.*

Global Variables

In the previous example, known is created outside the function, so it belongs to the special frame called __main__. Variables in __main__ are sometimes called **global** because they can be accessed from any function. Unlike local variables, which disappear when their function ends, global variables persist from one function call to the next.

It is common to use global variables for **flags**; that is, boolean variables that indicate ("flag") whether a condition is true. For example, some programs use a flag named verbose to control the level of detail in the output:

```
verbose = True

def example1():
    if verbose:
        print 'Running example1'
```

If you try to reassign a global variable, you might be surprised. The following example is supposed to keep track of whether the function has been called:

```
been_called = False

def example2():
    been_called = True        # WRONG
```

But if you run it you will see that the value of been_called doesn't change. The problem is that example2 creates a new local variable named been_called. The local variable goes away when the function ends, and has no effect on the global variable.

To reassign a global variable inside a function you have to **declare** the global variable before you use it:

```
been_called = False

def example2():
    global been_called
    been_called = True
```

The global statement tells the interpreter something like, "In this function, when I say been_called, I mean the global variable; don't create a local one."

Here's an example that tries to update a global variable:

```
count = 0

def example3():
    count = count + 1          # WRONG
```

If you run it you get:

```
UnboundLocalError: local variable 'count' referenced before assignment
```

Python assumes that count is local, which means that you are reading it before writing it. The solution, again, is to declare count global.

```
def example3():
    global count
    count += 1
```

If the global value is mutable, you can modify it without declaring it:

```
known = {0:0, 1:1}

def example4():
    known[2] = 1
```

So you can add, remove and replace elements of a global list or dictionary, but if you want to reassign the variable, you have to declare it:

```
def example5():
    global known
    known = dict()
```

Long Integers

If you compute fibonacci(50), you get:

```
>>> fibonacci(50)
12586269025L
```

The L at the end indicates that the result is a long integer, or type `long`. In Python 3, `long` is gone; all integers, even really big ones, are type `int`.

Values with type `int` have a limited range; long integers can be arbitrarily big, but as they get bigger they consume more space and time.

The mathematical operators work on long integers, and the functions in the `math` module, too, so in general any code that works with `int` will also work with `long`.

Any time the result of a computation is too big to be represented with an integer, Python converts the result as a long integer:

```
>>> 1000 * 1000
1000000
>>> 100000 * 100000
10000000000L
```

In the first case the result has type `int`; in the second case it is `long`.

Exercise 11-8.
Exponentiation of large integers is the basis of common algorithms for public-key encryption. Read the Wikipedia page on the RSA algorithm (*http://en.wikipedia.org/wiki/RSA*) and write functions to encode and decode messages.

Debugging

As you work with bigger datasets it can become unwieldy to debug by printing and checking data by hand. Here are some suggestions for debugging large datasets:

Scale down the input:
> If possible, reduce the size of the dataset. For example if the program reads a text file, start with just the first 10 lines, or with the smallest example you can find. You can either edit the files themselves, or (better) modify the program so it reads only the first n lines.
>
> If there is an error, you can reduce n to the smallest value that manifests the error, and then increase it gradually as you find and correct errors.

Check summaries and types:
> Instead of printing and checking the entire dataset, consider printing summaries of the data: for example, the number of items in a dictionary or the total of a list of numbers.
>
> A common cause of runtime errors is a value that is not the right type. For debugging this kind of error, it is often enough to print the type of a value.

Write self-checks:

Sometimes you can write code to check for errors automatically. For example, if you are computing the average of a list of numbers, you could check that the result is not greater than the largest element in the list or less than the smallest. This is called a "sanity check" because it detects results that are "insane."

Another kind of check compares the results of two different computations to see if they are consistent. This is called a "consistency check."

Pretty print the output:

Formatting debugging output can make it easier to spot an error. We saw an example in "Debugging" (page 70). The pprint module provides a pprint function that displays built-in types in a more human-readable format.

Again, time you spend building scaffolding can reduce the time you spend debugging.

Glossary

Dictionary:

A mapping from a set of keys to their corresponding values.

Key-value pair:

The representation of the mapping from a key to a value.

Item:

Another name for a key-value pair.

Key:

An object that appears in a dictionary as the first part of a key-value pair.

Value:

An object that appears in a dictionary as the second part of a key-value pair. This is more specific than our previous use of the word "value."

Implementation:

A way of performing a computation.

Hashtable:

The algorithm used to implement Python dictionaries.

Hash function:

A function used by a hashtable to compute the location for a key.

Hashable:

A type that has a hash function. Immutable types like integers, floats and strings are hashable; mutable types like lists and dictionaries are not.

Lookup:

A dictionary operation that takes a key and finds the corresponding value.

Reverse lookup:
>A dictionary operation that takes a value and finds one or more keys that map to it.

Singleton:
>A list (or other sequence) with a single element.

Call graph:
>A diagram that shows every frame created during the execution of a program, with an arrow from each caller to each callee.

Histogram:
>A set of counters.

Memo:
>A computed value stored to avoid unnecessary future computation.

Global variable:
>A variable defined outside a function. Global variables can be accessed from any function.

Flag:
>A boolean variable used to indicate whether a condition is true.

Declaration:
>A statement like global that tells the interpreter something about a variable.

Exercises

Exercise 11-9.
If you did Exercise 10-8, you already have a function named has_duplicates that takes a list as a parameter and returns True if there is any object that appears more than once in the list.

Use a dictionary to write a faster, simpler version of has_duplicates. Solution: *http://thinkpython.com/code/has_duplicates.py.*

Exercise 11-10.
Two words are "rotate pairs" if you can rotate one of them and get the other (see rotate_word in Exercise 8-12).

Write a program that reads a wordlist and finds all the rotate pairs. Solution: *http://thinkpython.com/code/rotate_pairs.py.*

Exercise 11-11.
Here's another Puzzler from *Car Talk* (*http://www.cartalk.com/content/puzzler/transcripts/200717*):

This was sent in by a fellow named Dan O'Leary. He came upon a common one-syllable, five-letter word recently that has the following unique property. When you remove the first letter, the remaining letters form a homophone of the original word, that is a word that sounds exactly the same. Replace the first letter, that is, put it back and remove the second letter and the result is yet another homophone of the original word. And the question is, what's the word?

Now I'm going to give you an example that doesn't work. Let's look at the five-letter word, 'wrack.' W-R-A-C-K, you know like to 'wrack with pain.' If I remove the first letter, I am left with a four-letter word, 'R-A-C-K.' As in, 'Holy cow, did you see the rack on that buck! It must have been a nine-pointer!' It's a perfect homophone. If you put the 'w' back, and remove the 'r,' instead, you're left with the word, 'wack,' which is a real word, it's just not a homophone of the other two words.

But there is, however, at least one word that Dan and we know of, which will yield two homophones if you remove either of the first two letters to make two, new four-letter words. The question is, what's the word?

You can use the dictionary from Exercise 11-1 to check whether a string is in the word list.

To check whether two words are homophones, you can use the CMU Pronouncing Dictionary. You can download it from here (*http://www.speech.cs.cmu.edu/cgi-bin/ cmudict*) or from here (*http://thinkpython.com/code/c06d*), and you can also download *http://thinkpython.com/code/pronounce.py*, which provides a function named read_dictionary that reads the pronouncing dictionary and returns a Python dictionary that maps from each word to a string that describes its primary pronunciation.

Write a program that lists all the words that solve the Puzzler. Solution: *http:// thinkpython.com/code/homophone.py*.

## CHAPTER 12

Tuples

Tuples Are Immutable

A tuple is a sequence of values. The values can be any type, and they are indexed by integers, so in that respect tuples are a lot like lists. The important difference is that tuples are immutable.

Syntactically, a tuple is a comma-separated list of values:

```
>>> t = 'a', 'b', 'c', 'd', 'e'
```

Although it is not necessary, it is common to enclose tuples in parentheses:

```
>>> t = ('a', 'b', 'c', 'd', 'e')
```

To create a tuple with a single element, you have to include a final comma:

```
>>> t1 = 'a',
>>> type(t1)
<type 'tuple'>
```

A value in parentheses is not a tuple:

```
>>> t2 = ('a')
>>> type(t2)
<type 'str'>
```

Another way to create a tuple is the built-in function tuple. With no argument, it creates an empty tuple:

```
>>> t = tuple()
>>> print t
()
```

If the argument is a sequence (string, list or tuple), the result is a tuple with the elements of the sequence:

```
>>> t = tuple('lupins')
>>> print t
('l', 'u', 'p', 'i', 'n', 's')
```

Because `tuple` is the name of a built-in function, you should avoid using it as a variable name.

Most list operators also work on tuples. The bracket operator indexes an element:

```
>>> t = ('a', 'b', 'c', 'd', 'e')
>>> print t[0]
'a'
```

And the slice operator selects a range of elements.

```
>>> print t[1:3]
('b', 'c')
```

But if you try to modify one of the elements of the tuple, you get an error:

```
>>> t[0] = 'A'
TypeError: object doesn't support item assignment
```

You can't modify the elements of a tuple, but you can replace one tuple with another:

```
>>> t = ('A',) + t[1:]
>>> print t
('A', 'b', 'c', 'd', 'e')
```

Tuple Assignment

It is often useful to swap the values of two variables. With conventional assignments, you have to use a temporary variable. For example, to swap a and b:

```
>>> temp = a
>>> a = b
>>> b = temp
```

This solution is cumbersome; **tuple assignment** is more elegant:

```
>>> a, b = b, a
```

The left side is a tuple of variables; the right side is a tuple of expressions. Each value is assigned to its respective variable. All the expressions on the right side are evaluated before any of the assignments.

The number of variables on the left and the number of values on the right have to be the same:

```
>>> a, b = 1, 2, 3
ValueError: too many values to unpack
```

More generally, the right side can be any kind of sequence (string, list or tuple). For example, to split an email address into a user name and a domain, you could write:

```
>>> addr = 'monty@python.org'
>>> uname, domain = addr.split('@')
```

The return value from split is a list with two elements; the first element is assigned to uname, the second to domain.

```
>>> print uname
monty
>>> print domain
python.org
```

Tuples as Return Values

Strictly speaking, a function can only return one value, but if the value is a tuple, the effect is the same as returning multiple values. For example, if you want to divide two integers and compute the quotient and remainder, it is inefficient to compute x/y and then x%y. It is better to compute them both at the same time.

The built-in function divmod takes two arguments and returns a tuple of two values, the quotient and remainder. You can store the result as a tuple:

```
>>> t = divmod(7, 3)
>>> print t
(2, 1)
```

Or use tuple assignment to store the elements separately:

```
>>> quot, rem = divmod(7, 3)
>>> print quot
2
>>> print rem
1
```

Here is an example of a function that returns a tuple:

```
def min_max(t):
    return min(t), max(t)
```

max and min are built-in functions that find the largest and smallest elements of a sequence. min_max computes both and returns a tuple of two values.

Variable-Length Argument Tuples

Functions can take a variable number of arguments. A parameter name that begins with * **gathers** arguments into a tuple. For example, printall takes any number of arguments and prints them:

```
def printall(*args):
    print args
```

The gather parameter can have any name you like, but args is conventional. Here's how the function works:

```
>>> printall(1, 2.0, '3')
(1, 2.0, '3')
```

The complement of gather is **scatter**. If you have a sequence of values and you want to pass it to a function as multiple arguments, you can use the * operator. For example, divmod takes exactly two arguments; it doesn't work with a tuple:

```
>>> t = (7, 3)
>>> divmod(t)
TypeError: divmod expected 2 arguments, got 1
```

But if you scatter the tuple, it works:

```
>>> divmod(*t)
(2, 1)
```

Exercise 12-1.

Many of the built-in functions use variable-length argument tuples. For example, max and min can take any number of arguments:

```
>>> max(1,2,3)
3
```

But sum does not.

```
>>> sum(1,2,3)
TypeError: sum expected at most 2 arguments, got 3
```

Write a function called sumall that takes any number of arguments and returns their sum.

Lists and Tuples

zip is a built-in function that takes two or more sequences and "zips" them into a list of tuples where each tuple contains one element from each sequence. In Python 3, zip returns an iterator of tuples, but for most purposes, an iterator behaves like a list.

This example zips a string and a list:

```
>>> s = 'abc'
>>> t = [0, 1, 2]
>>> zip(s, t)
[('a', 0), ('b', 1), ('c', 2)]
```

The result is a list of tuples where each tuple contains a character from the string and the corresponding element from the list.

If the sequences are not the same length, the result has the length of the shorter one.

```
>>> zip('Anne', 'Elk')
[('A', 'E'), ('n', 'l'), ('n', 'k')]
```

You can use tuple assignment in a for loop to traverse a list of tuples:

```
t = [('a', 0), ('b', 1), ('c', 2)]
for letter, number in t:
    print number, letter
```

Each time through the loop, Python selects the next tuple in the list and assigns the elements to letter and number. The output of this loop is:

```
0 a
1 b
2 c
```

If you combine zip, for and tuple assignment, you get a useful idiom for traversing two (or more) sequences at the same time. For example, has_match takes two sequences, t1 and t2, and returns True if there is an index i such that t1[i] == t2[i]:

```
def has_match(t1, t2):
    for x, y in zip(t1, t2):
        if x == y:
            return True
    return False
```

If you need to traverse the elements of a sequence and their indices, you can use the built-in function enumerate:

```
for index, element in enumerate('abc'):
    print index, element
```

The output of this loop is:

```
0 a
1 b
2 c
```

Again.

Dictionaries and Tuples

Dictionaries have a method called items that returns a list of tuples, where each tuple is a key-value pair.

```
>>> d = {'a':0, 'b':1, 'c':2}
>>> t = d.items()
>>> print t
[('a', 0), ('c', 2), ('b', 1)]
```

As you should expect from a dictionary, the items are in no particular order. In Python 3, items returns an iterator, but for many purposes, iterators behave like lists.

Going in the other direction, you can use a list of tuples to initialize a new dictionary:

```
>>> t = [('a', 0), ('c', 2), ('b', 1)]
>>> d = dict(t)
>>> print d
{'a': 0, 'c': 2, 'b': 1}
```

Combining `dict` with `zip` yields a concise way to create a dictionary:

```
>>> d = dict(zip('abc', range(3)))
>>> print d
{'a': 0, 'c': 2, 'b': 1}
```

The dictionary method `update` also takes a list of tuples and adds them, as key-value pairs, to an existing dictionary.

Combining `items`, tuple assignment and `for`, you get the idiom for traversing the keys and values of a dictionary:

```
for key, val in d.items():
    print val, key
```

The output of this loop is:

```
0 a
2 c
1 b
```

Again.

It is common to use tuples as keys in dictionaries (primarily because you can't use lists). For example, a telephone directory might map from last-name, first-name pairs to telephone numbers. Assuming that we have defined `last`, `first` and `number`, we could write:

```
directory[last,first] = number
```

The expression in brackets is a tuple. We could use tuple assignment to traverse this dictionary.

```
for last, first in directory:
    print first, last, directory[last,first]
```

This loop traverses the keys in `directory`, which are tuples. It assigns the elements of each tuple to `last` and `first`, then prints the name and corresponding telephone number.

There are two ways to represent tuples in a state diagram. The more detailed version shows the indices and elements just as they appear in a list. For example, the tuple (`'Cleese'`, `'John'`) would appear as in Figure 12-1.

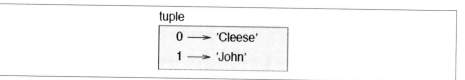

Figure 12-1. State diagram.

But in a larger diagram you might want to leave out the details. For example, a diagram of the telephone directory might appear as in Figure 12-2.

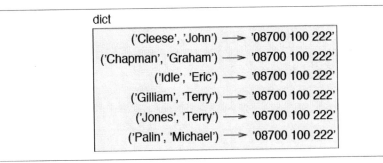

Figure 12-2. State diagram.

Here the tuples are shown using Python syntax as a graphical shorthand.

The telephone number in the diagram is the complaints line for the BBC, so please don't call it.

Comparing Tuples

The relational operators work with tuples and other sequences; Python starts by comparing the first element from each sequence. If they are equal, it goes on to the next elements, and so on, until it finds elements that differ. Subsequent elements are not considered (even if they are really big).

```
>>> (0, 1, 2) < (0, 3, 4)
True
>>> (0, 1, 2000000) < (0, 3, 4)
True
```

The sort function works the same way. It sorts primarily by first element, but in the case of a tie, it sorts by second element, and so on.

This feature lends itself to a pattern called **DSU** for:

Decorate
 a sequence by building a list of tuples with one or more sort keys preceding the elements from the sequence,

Sort
 the list of tuples, and

Undecorate
 by extracting the sorted elements of the sequence.

For example, suppose you have a list of words and you want to sort them from longest to shortest:

```
def sort_by_length(words):
    t = []
    for word in words:
        t.append((len(word), word))

    t.sort(reverse=True)

    res = []
    for length, word in t:
        res.append(word)
    return res
```

The first loop builds a list of tuples, where each tuple is a word preceded by its length.

`sort` compares the first element, length, first, and only considers the second element to break ties. The keyword argument `reverse=True` tells `sort` to go in decreasing order.

The second loop traverses the list of tuples and builds a list of words in descending order of length.

Exercise 12-2.

In this example, ties are broken by comparing words, so words with the same length appear in reverse alphabetical order. For other applications you might want to break ties at random. Modify this example so that words with the same length appear in random order. Hint: see the `random` function in the `random` module. Solution: *http://thinkpython.com/code/unstable_sort.py.*

Sequences of Sequences

I have focused on lists of tuples, but almost all of the examples in this chapter also work with lists of lists, tuples of tuples, and tuples of lists. To avoid enumerating the possible combinations, it is sometimes easier to talk about sequences of sequences.

In many contexts, the different kinds of sequences (strings, lists and tuples) can be used interchangeably. So how and why do you choose one over the others?

To start with the obvious, strings are more limited than other sequences because the elements have to be characters. They are also immutable. If you need the ability to change the characters in a string (as opposed to creating a new string), you might want to use a list of characters instead.

Lists are more common than tuples, mostly because they are mutable. But there are a few cases where you might prefer tuples:

1. In some contexts, like a `return` statement, it is syntactically simpler to create a tuple than a list. In other contexts, you might prefer a list.

2. If you want to use a sequence as a dictionary key, you have to use an immutable type like a tuple or string.

3. If you are passing a sequence as an argument to a function, using tuples reduces the potential for unexpected behavior due to aliasing.

Because tuples are immutable, they don't provide methods like sort and reverse, which modify existing lists. But Python provides the built-in functions sorted and reversed, which take any sequence as a parameter and return a new list with the same elements in a different order.

Debugging

Lists, dictionaries and tuples are known generically as **data structures**; in this chapter we are starting to see compound data structures, like lists of tuples, and dictionaries that contain tuples as keys and lists as values. Compound data structures are useful, but they are prone to what I call **shape errors**; that is, errors caused when a data structure has the wrong type, size or composition. For example, if you are expecting a list with one integer and I give you a plain old integer (not in a list), it won't work.

To help debug these kinds of errors, I have written a module called structshape that provides a function, also called structshape, that takes any kind of data structure as an argument and returns a string that summarizes its shape. You can download it from *http://thinkpython.com/code/structshape.py*

Here's the result for a simple list:

```
>>> from structshape import structshape
>>> t = [1,2,3]
>>> print structshape(t)
list of 3 int
```

A fancier program might write "list of 3 int*s*," but it was easier not to deal with plurals. Here's a list of lists:

```
>>> t2 = [[1,2], [3,4], [5,6]]
>>> print structshape(t2)
list of 3 list of 2 int
```

If the elements of the list are not the same type, structshape groups them, in order, by type:

```
>>> t3 = [1, 2, 3, 4.0, '5', '6', [7], [8], 9]
>>> print structshape(t3)
list of (3 int, float, 2 str, 2 list of int, int)
```

Here's a list of tuples:

```
>>> s = 'abc'
>>> lt = zip(t, s)
>>> print structshape(lt)
list of 3 tuple of (int, str)
```

And here's a dictionary with 3 items that map integers to strings.

```
>>> d = dict(lt)
>>> print structshape(d)
dict of 3 int->str
```

If you are having trouble keeping track of your data structures, structshape can help.

Glossary

Tuple:
An immutable sequence of elements.

Tuple assignment:
An assignment with a sequence on the right side and a tuple of variables on the left. The right side is evaluated and then its elements are assigned to the variables on the left.

Gather:
The operation of assembling a variable-length argument tuple.

Scatter:
The operation of treating a sequence as a list of arguments.

DSU:
Abbreviation of "decorate-sort-undecorate," a pattern that involves building a list of tuples, sorting, and extracting part of the result.

Data structure:
A collection of related values, often organized in lists, dictionaries, tuples, etc.

Shape (of a data structure):
A summary of the type, size and composition of a data structure.

Exercises

Exercise 12-3.
Write a function called most_frequent that takes a string and prints the letters in decreasing order of frequency. Find text samples from several different languages and see how letter frequency varies between languages. Compare your results with the tables at *http://en.wikipedia.org/wiki/Letter_frequencies*. Solution: *http://thinkpython.com/code/most_frequent.py*.

Exercise 12-4.
More anagrams!

1. Write a program that reads a word list from a file (see "Reading Word Lists" (page 97)) and prints all the sets of words that are anagrams.

   Here is an example of what the output might look like:

   ```
   ['deltas', 'desalt', 'lasted', 'salted', 'slated', 'staled']
   ['retainers', 'ternaries']
   ['generating', 'greatening']
   ['resmelts', 'smelters', 'termless']
   ```

   Hint: you might want to build a dictionary that maps from a set of letters to a list of words that can be spelled with those letters. The question is, how can you represent the set of letters in a way that can be used as a key?

2. Modify the previous program so that it prints the largest set of anagrams first, followed by the second largest set, and so on.

3. In Scrabble a "bingo" is when you play all seven tiles in your rack, along with a letter on the board, to form an eight-letter word. What set of 8 letters forms the most possible bingos? Hint: there are seven.

   Solution: *http://thinkpython.com/code/anagram_sets.py.*

Exercise 12-5.
Two words form a "metathesis pair" if you can transform one into the other by swapping two letters; for example, "converse" and "conserve." Write a program that finds all of the metathesis pairs in the dictionary. Hint: don't test all pairs of words, and don't test all possible swaps. Solution: *http://thinkpython.com/code/metathesis.py.* Credit: This exercise is inspired by an example at *http://puzzlers.org.*

Exercise 12-6.
Here's another Car Talk Puzzler (*http://www.cartalk.com/content/puzzler/transcripts/200651*):

> What is the longest English word, that remains a valid English word, as you remove its letters one at a time?
>
> Now, letters can be removed from either end, or the middle, but you can't rearrange any of the letters. Every time you drop a letter, you wind up with another English word. If you do that, you're eventually going to wind up with one letter and that too is going to be an English word—one that's found in the dictionary. I want to know what's the longest word and how many letters does it have?
>
> I'm going to give you a little modest example: Sprite. Ok? You start off with sprite, you take a letter off, one from the interior of the word, take the r away, and we're left with the word spite, then we take the e off the end, we're left with spit, we take the s off, we're left with pit, it, and I.

Write a program to find all words that can be reduced in this way, and then find the longest one.

This exercise is a little more challenging than most, so here are some suggestions:

1. You might want to write a function that takes a word and computes a list of all the words that can be formed by removing one letter. These are the "children" of the word.

2. Recursively, a word is reducible if any of its children are reducible. As a base case, you can consider the empty string reducible.

3. The wordlist I provided, words.txt, doesn't contain single letter words. So you might want to add "I", "a", and the empty string.

4. To improve the performance of your program, you might want to memoize the words that are known to be reducible.

Solution: *http://thinkpython.com/code/reducible.py.*

## CHAPTER 13

Case Study: Data Structure Selection

Word Frequency Analysis

As usual, you should at least attempt the following exercises before you read my solutions.

Exercise 13-1.
Write a program that reads a file, breaks each line into words, strips whitespace and punctuation from the words, and converts them to lowercase.

Hint: The `string` module provides strings named `whitespace`, which contains space, tab, newline, etc., and `punctuation` which contains the punctuation characters. Let's see if we can make Python swear:

```
>>> import string
>>> print string.punctuation
!"#$%&'()*+,-./:;<=>?@[\]^_`{|}~
```

Also, you might consider using the string methods `strip`, `replace` and `translate`.

Exercise 13-2.
Go to Project Gutenberg (*http://www.gutenberg.org*) and download your favorite out-of-copyright book in plain text format.

Modify your program from the previous exercise to read the book you downloaded, skip over the header information at the beginning of the file, and process the rest of the words as before.

Then modify the program to count the total number of words in the book, and the number of times each word is used.

Print the number of different words used in the book. Compare different books by different authors, written in different eras. Which author uses the most extensive vocabulary?

Exercise 13-3.
Modify the program from the previous exercise to print the 20 most frequently-used words in the book.

Exercise 13-4.
Modify the previous program to read a word list (see "Reading Word Lists" (page 97)) and then print all the words in the book that are not in the word list. How many of them are typos? How many of them are common words that *should* be in the word list, and how many of them are really obscure?

Random Numbers

Given the same inputs, most computer programs generate the same outputs every time, so they are said to be **deterministic**. Determinism is usually a good thing, since we expect the same calculation to yield the same result. For some applications, though, we want the computer to be unpredictable. Games are an obvious example, but there are more.

Making a program truly nondeterministic turns out to be not so easy, but there are ways to make it at least seem nondeterministic. One of them is to use algorithms that generate **pseudorandom** numbers. Pseudorandom numbers are not truly random because they are generated by a deterministic computation, but just by looking at the numbers it is all but impossible to distinguish them from random.

The random module provides functions that generate pseudorandom numbers (which I will simply call "random" from here on).

The function random returns a random float between 0.0 and 1.0 (including 0.0 but not 1.0). Each time you call random, you get the next number in a long series. To see a sample, run this loop:

```
import random

for i in range(10):
    x = random.random()
    print x
```

The function randint takes parameters low and high and returns an integer between low and high (including both).

```
>>> random.randint(5, 10)
5
>>> random.randint(5, 10)
9
```

To choose an element from a sequence at random, you can use `choice`:

```
>>> t = [1, 2, 3]
>>> random.choice(t)
2
>>> random.choice(t)
3
```

The `random` module also provides functions to generate random values from continuous distributions including Gaussian, exponential, gamma, and a few more.

Exercise 13-5.
Write a function named `choose_from_hist` that takes a histogram as defined in "Dictionary as a Set of Counters" (page 123) and returns a random value from the histogram, chosen with probability in proportion to frequency. For example, for this histogram:

```
>>> t = ['a', 'a', 'b']
>>> hist = histogram(t)
>>> print hist
{'a': 2, 'b': 1}
```

your function should return 'a' with probability *2/3* and 'b' with probability *1/3*.

Word Histogram

You should attempt the previous exercises before you go on. You can download my solution from *http://thinkpython.com/code/analyze_book.py*. You will also need *http://thinkpython.com/code/emma.txt*.

Here is a program that reads a file and builds a histogram of the words in the file:

```
import string

def process_file(filename):
    hist = dict()
    fp = open(filename)
    for line in fp:
        process_line(line, hist)
    return hist

def process_line(line, hist):
    line = line.replace('-', ' ')

    for word in line.split():
        word = word.strip(string.punctuation + string.whitespace)
        word = word.lower()

        hist[word] = hist.get(word, 0) + 1

hist = process_file('emma.txt')
```

This program reads `emma.txt`, which contains the text of *Emma* by Jane Austen.

`process_file` loops through the lines of the file, passing them one at a time to `process_line`. The histogram `hist` is being used as an accumulator.

`process_line` uses the string method `replace` to replace hyphens with spaces before using `split` to break the line into a list of strings. It traverses the list of words and uses `strip` and `lower` to remove punctuation and convert to lower case. (It is a shorthand to say that strings are "converted;" remember that string are immutable, so methods like `strip` and `lower` return new strings.)

Finally, `process_line` updates the histogram by creating a new item or incrementing an existing one.

To count the total number of words in the file, we can add up the frequencies in the histogram:

```
def total_words(hist):
    return sum(hist.values())
```

The number of different words is just the number of items in the dictionary:

```
def different_words(hist):
    return len(hist)
```

Here is some code to print the results:

```
print 'Total number of words:', total_words(hist)
print 'Number of different words:', different_words(hist)
```

And the results:

```
Total number of words: 161080
Number of different words: 7214
```

Most Common Words

To find the most common words, we can apply the DSU pattern; `most_common` takes a histogram and returns a list of word-frequency tuples, sorted in reverse order by frequency:

```
def most_common(hist):
    t = []
    for key, value in hist.items():
        t.append((value, key))

    t.sort(reverse=True)
    return t
```

Here is a loop that prints the ten most common words:

```
t = most_common(hist)
print 'The most common words are:'
for freq, word in t[0:10]:
    print word, '\t', freq
```

And here are the results from *Emma*:

```
The most common words are:
to   5242
the  5205
and  4897
of   4295
i    3191
a    3130
it   2529
her  2483
was  2400
she  2364
```

Optional Parameters

We have seen built-in functions and methods that take a variable number of arguments. It is possible to write user-defined functions with optional arguments, too. For example, here is a function that prints the most common words in a histogram

```
def print_most_common(hist, num=10):
    t = most_common(hist)
    print 'The most common words are:'
    for freq, word in t[:num]:
        print word, '\t', freq
```

The first parameter is required; the second is optional. The **default value** of num is 10.

If you only provide one argument:

```
print_most_common(hist)
```

num gets the default value. If you provide two arguments:

```
print_most_common(hist, 20)
```

num gets the value of the argument instead. In other words, the optional argument **overrides** the default value.

If a function has both required and optional parameters, all the required parameters have to come first, followed by the optional ones.

Dictionary Subtraction

Finding the words from the book that are not in the word list from words.txt is a problem you might recognize as set subtraction; that is, we want to find all the words from one set (the words in the book) that are not in another set (the words in the list).

subtract takes dictionaries d1 and d2 and returns a new dictionary that contains all the keys from d1 that are not in d2. Since we don't really care about the values, we set them all to None.

```
def subtract(d1, d2):
    res = dict()
    for key in d1:
        if key not in d2:
            res[key] = None
    return res
```

To find the words in the book that are not in `words.txt`, we can use `process_file` to build a histogram for `words.txt`, and then subtract:

```
words = process_file('words.txt')
diff = subtract(hist, words)

print "The words in the book that aren't in the word list are:"
for word in diff.keys():
    print word,
```

Here are some of the results from *Emma*:

```
The words in the book that aren't in the word list are:
 rencontre jane's blanche woodhouses disingenuousness
friend's venice apartment ...
```

Some of these words are names and possessives. Others, like "rencontre," are no longer in common use. But a few are common words that should really be in the list!

Exercise 13-6.

Python provides a data structure called `set` that provides many common set operations. Read the documentation at *http://docs.python.org/lib/types-set.html* and write a program that uses set subtraction to find words in the book that are not in the word list. Solution: *http://thinkpython.com/code/analyze_book2.py*.

Random Words

To choose a random word from the histogram, the simplest algorithm is to build a list with multiple copies of each word, according to the observed frequency, and then choose from the list:

```
def random_word(h):
    t = []
    for word, freq in h.items():
        t.extend([word] * freq)

    return random.choice(t)
```

The expression `[word] * freq` creates a list with `freq` copies of the string `word`. The `extend` method is similar to `append` except that the argument is a sequence.

Exercise 13-7.

This algorithm works, but it is not very efficient; each time you choose a random word, it rebuilds the list, which is as big as the original book. An obvious improvement is to build the list once and then make multiple selections, but the list is still big.

An alternative is:

1. Use keys to get a list of the words in the book.
2. Build a list that contains the cumulative sum of the word frequencies (see Exercise 10-3). The last item in this list is the total number of words in the book, *n*.
3. Choose a random number from 1 to *n*. Use a bisection search (See Exercise 10-11) to find the index where the random number would be inserted in the cumulative sum.
4. Use the index to find the corresponding word in the word list.

Write a program that uses this algorithm to choose a random word from the book. Solution: *http://thinkpython.com/code/analyze_book3.py.*

Markov Analysis

If you choose words from the book at random, you can get a sense of the vocabulary, you probably won't get a sentence:

```
this the small regard harriet which knightley's it most things
```

A series of random words seldom makes sense because there is no relationship between successive words. For example, in a real sentence you would expect an article like "the" to be followed by an adjective or a noun, and probably not a verb or adverb.

One way to measure these kinds of relationships is Markov analysis, which characterizes, for a given sequence of words, the probability of the word that comes next. For example, the song *Eric, the Half a Bee* begins:

> Half a bee, philosophically, Must, ipso facto, half not be. But half the bee has got to be Vis a vis, its entity. D'you see? But can a bee be said to be Or not to be an entire bee When half the bee is not a bee Due to some ancient injury?

In this text, the phrase "half the" is always followed by the word "bee," but the phrase "the bee" might be followed by either "has" or "is."

The result of Markov analysis is a mapping from each prefix (like "half the" and "the bee") to all possible suffixes (like "has" and "is").

Given this mapping, you can generate a random text by starting with any prefix and choosing at random from the possible suffixes. Next, you can combine the end of the prefix and the new suffix to form the next prefix, and repeat.

For example, if you start with the prefix "Half a," then the next word has to be "bee," because the prefix only appears once in the text. The next prefix is "a bee," so the next suffix might be "philosophically," "be" or "due."

In this example the length of the prefix is always two, but you can do Markov analysis with any prefix length. The length of the prefix is called the "order" of the analysis.

Exercise 13-8.
Markov analysis:

1. Write a program to read a text from a file and perform Markov analysis. The result should be a dictionary that maps from prefixes to a collection of possible suffixes. The collection might be a list, tuple, or dictionary; it is up to you to make an appropriate choice. You can test your program with prefix length two, but you should write the program in a way that makes it easy to try other lengths.

2. Add a function to the previous program to generate random text based on the Markov analysis. Here is an example from *Emma* with prefix length 2:

> He was very clever, be it sweetness or be angry, ashamed or only amused, at such a stroke. She had never thought of Hannah till you were never meant for me?" "I cannot make speeches, Emma:" he soon cut it all himself.

For this example, I left the punctuation attached to the words. The result is almost syntactically correct, but not quite. Semantically, it almost makes sense, but not quite.

What happens if you increase the prefix length? Does the random text make more sense?

3. Once your program is working, you might want to try a mash-up: if you analyze text from two or more books, the random text you generate will blend the vocabulary and phrases from the sources in interesting ways.

Credit: This case study is based on an example from Kernighan and Pike, *The Practice of Programming*, Addison-Wesley, 1999.

You should attempt this exercise before you go on; then you can can download my solution from *http://thinkpython.com/code/markov.py*. You will also need *http://thinkpython.com/code/emma.txt*.

Data Structures

Using Markov analysis to generate random text is fun, but there is also a point to this exercise: data structure selection. In your solution to the previous exercises, you had to choose:

- How to represent the prefixes.

- How to represent the collection of possible suffixes.

- How to represent the mapping from each prefix to the collection of possible suffixes.

Ok, the last one is the easy; the only mapping type we have seen is a dictionary, so it is the natural choice.

For the prefixes, the most obvious options are string, list of strings, or tuple of strings. For the suffixes, one option is a list; another is a histogram (dictionary).

How should you choose? The first step is to think about the operations you will need to implement for each data structure. For the prefixes, we need to be able to remove words from the beginning and add to the end. For example, if the current prefix is "Half a," and the next word is "bee," you need to be able to form the next prefix, "a bee."

Your first choice might be a list, since it is easy to add and remove elements, but we also need to be able to use the prefixes as keys in a dictionary, so that rules out lists. With tuples, you can't append or remove, but you can use the addition operator to form a new tuple:

```
def shift(prefix, word):
    return prefix[1:] + (word,)
```

shift takes a tuple of words, prefix, and a string, word, and forms a new tuple that has all the words in prefix except the first, and word added to the end.

For the collection of suffixes, the operations we need to perform include adding a new suffix (or increasing the frequency of an existing one), and choosing a random suffix.

Adding a new suffix is equally easy for the list implementation or the histogram. Choosing a random element from a list is easy; choosing from a histogram is harder to do efficiently (see Exercise 13-7).

So far we have been talking mostly about ease of implementation, but there are other factors to consider in choosing data structures. One is run time. Sometimes there is a theoretical reason to expect one data structure to be faster than other; for example, I mentioned that the in operator is faster for dictionaries than for lists, at least when the number of elements is large.

But often you don't know ahead of time which implementation will be faster. One option is to implement both of them and see which is better. This approach is called **benchmarking**. A practical alternative is to choose the data structure that is easiest to implement, and then see if it is fast enough for the intended application. If so, there is no need to go on. If not, there are tools, like the profile module, that can identify the places in a program that take the most time.

The other factor to consider is storage space. For example, using a histogram for the collection of suffixes might take less space because you only have to store each word once, no matter how many times it appears in the text. In some cases, saving space can also make your program run faster, and in the extreme, your program might not run at all if you run out of memory. But for many applications, space is a secondary consideration after run time.

One final thought: in this discussion, I have implied that we should use one data structure for both analysis and generation. But since these are separate phases, it would also be possible to use one structure for analysis and then convert to another structure for generation. This would be a net win if the time saved during generation exceeded the time spent in conversion.

Debugging

When you are debugging a program, and especially if you are working on a hard bug, there are four things to try:

Reading:
> Examine your code, read it back to yourself, and check that it says what you meant to say.

Running:
> Experiment by making changes and running different versions. Often if you display the right thing at the right place in the program, the problem becomes obvious, but sometimes you have to spend some time to build scaffolding.

Ruminating:
> Take some time to think! What kind of error is it: syntax, runtime, semantic? What information can you get from the error messages, or from the output of the program? What kind of error could cause the problem you're seeing? What did you change last, before the problem appeared?

Retreating:
> At some point, the best thing to do is back off, undoing recent changes, until you get back to a program that works and that you understand. Then you can start rebuilding.

Beginning programmers sometimes get stuck on one of these activities and forget the others. Each activity comes with its own failure mode.

For example, reading your code might help if the problem is a typographical error, but not if the problem is a conceptual misunderstanding. If you don't understand what your program does, you can read it 100 times and never see the error, because the error is in your head.

Running experiments can help, especially if you run small, simple tests. But if you run experiments without thinking or reading your code, you might fall into a pattern I call "random walk programming," which is the process of making random changes until the program does the right thing. Needless to say, random walk programming can take a long time.

You have to take time to think. Debugging is like an experimental science. You should have at least one hypothesis about what the problem is. If there are two or more possibilities, try to think of a test that would eliminate one of them.

Taking a break helps with the thinking. So does talking. If you explain the problem to someone else (or even yourself), you will sometimes find the answer before you finish asking the question.

But even the best debugging techniques will fail if there are too many errors, or if the code you are trying to fix is too big and complicated. Sometimes the best option is to retreat, simplifying the program until you get to something that works and that you understand.

Beginning programmers are often reluctant to retreat because they can't stand to delete a line of code (even if it's wrong). If it makes you feel better, copy your program into another file before you start stripping it down. Then you can paste the pieces back in a little bit at a time.

Finding a hard bug requires reading, running, ruminating, and sometimes retreating. If you get stuck on one of these activities, try the others.

Glossary

Deterministic:
  Pertaining to a program that does the same thing each time it runs, given the same inputs.

Pseudorandom:
  Pertaining to a sequence of numbers that appear to be random, but are generated by a deterministic program.

Default value:
  The value given to an optional parameter if no argument is provided.

Override:
  To replace a default value with an argument.

Benchmarking:
  The process of choosing between data structures by implementing alternatives and testing them on a sample of the possible inputs.

Exercises

Exercise 13-9.

The "rank" of a word is its position in a list of words sorted by frequency: the most common word has rank 1, the second most common has rank 2, etc.

Zipf's law describes a relationship between the ranks and frequencies of words in natural languages (*http://en.wikipedia.org/wiki/Zipfs_law*). Specifically, it predicts that the frequency, f, of the word with rank r is:

$$f = cr^{-s}$$

where s and c are parameters that depend on the language and the text. If you take the logarithm of both sides of this equation, you get:

$$\log f = \log c - s \log r$$

So if you plot $\log f$ versus $\log r$, you should get a straight line with slope $-s$ and intercept $\log c$.

Write a program that reads a text from a file, counts word frequencies, and prints one line for each word, in descending order of frequency, with $\log f$ and $\log r$. Use the graphing program of your choice to plot the results and check whether they form a straight line. Can you estimate the value of s?

Solution: *http://thinkpython.com/code/zipf.py*. To make the plots, you might have to install matplotlib (see *http://matplotlib.sourceforge.net/*).

## CHAPTER 14

Files

Persistence

Most of the programs we have seen so far are transient in the sense that they run for a short time and produce some output, but when they end, their data disappears. If you run the program again, it starts with a clean slate.

Other programs are **persistent**: they run for a long time (or all the time); they keep at least some of their data in permanent storage (a hard drive, for example); and if they shut down and restart, they pick up where they left off.

Examples of persistent programs are operating systems, which run pretty much whenever a computer is on, and web servers, which run all the time, waiting for requests to come in on the network.

One of the simplest ways for programs to maintain their data is by reading and writing text files. We have already seen programs that read text files; in this chapter we will see programs that write them.

An alternative is to store the state of the program in a database. In this chapter I will present a simple database and a module, pickle, that makes it easy to store program data.

Reading and Writing

A text file is a sequence of characters stored on a permanent medium like a hard drive, flash memory, or CD-ROM. We saw how to open and read a file in "Reading Word Lists" (page 97).

To write a file, you have to open it with mode 'w' as a second parameter:

```
>>> fout = open('output.txt', 'w')
>>> print fout
<open file 'output.txt', mode 'w' at 0xb7eb2410>
```

If the file already exists, opening it in write mode clears out the old data and starts fresh, so be careful! If the file doesn't exist, a new one is created.

The `write` method puts data into the file.

```
>>> line1 = "This here's the wattle,\n"
>>> fout.write(line1)
```

Again, the file object keeps track of where it is, so if you call `write` again, it adds the new data to the end.

```
>>> line2 = "the emblem of our land.\n"
>>> fout.write(line2)
```

When you are done writing, you have to close the file.

```
>>> fout.close()
```

Format Operator

The argument of `write` has to be a string, so if we want to put other values in a file, we have to convert them to strings. The easiest way to do that is with `str`:

```
>>> x = 52
>>> f.write(str(x))
```

An alternative is to use the **format operator**, `%`. When applied to integers, `%` is the modulus operator. But when the first operand is a string, `%` is the format operator.

The first operand is the **format string**, which contains one or more **format sequences**, which specify how the second operand is formatted. The result is a string.

For example, the format sequence `'%d'` means that the second operand should be formatted as an integer (d stands for "decimal"):

```
>>> camels = 42
>>> '%d' % camels
'42'
```

The result is the string `'42'`, which is not to be confused with the integer value 42.

A format sequence can appear anywhere in the string, so you can embed a value in a sentence:

```
>>> camels = 42
>>> 'I have spotted %d camels.' % camels
'I have spotted 42 camels.'
```

If there is more than one format sequence in the string, the second argument has to be a tuple. Each format sequence is matched with an element of the tuple, in order.

The following example uses `'%d'` to format an integer, `'%g'` to format a floating-point number (don't ask why), and `'%s'` to format a string:

```
>>> 'In %d years I have spotted %g %s.' % (3, 0.1, 'camels')
'In 3 years I have spotted 0.1 camels.'
```

The number of elements in the tuple has to match the number of format sequences in the string. Also, the types of the elements have to match the format sequences:

```
>>> '%d %d %d' % (1, 2)
TypeError: not enough arguments for format string
>>> '%d' % 'dollars'
TypeError: illegal argument type for built-in operation
```

In the first example, there aren't enough elements; in the second, the element is the wrong type.

The format operator is powerful, but it can be difficult to use. You can read more about it at *docs.python.org/lib/typesseq-strings.html*.

Filenames and Paths

Files are organized into **directories** (also called "folders"). Every running program has a "current directory," which is the default directory for most operations. For example, when you open a file for reading, Python looks for it in the current directory.

The os module provides functions for working with files and directories ("os" stands for "operating system"). `os.getcwd` returns the name of the current directory:

```
>>> import os
>>> cwd = os.getcwd()
>>> print cwd
/home/dinsdale
```

cwd stands for "current working directory." The result in this example is `/home/dins dale`, which is the home directory of a user named dinsdale.

A string like cwd that identifies a file is called a **path**. A **relative path** starts from the current directory; an **absolute path** starts from the topmost directory in the file system.

The paths we have seen so far are simple filenames, so they are relative to the current directory. To find the absolute path to a file, you can use `os.path.abspath`:

```
>>> os.path.abspath('memo.txt')
'/home/dinsdale/memo.txt'
```

`os.path.exists` checks whether a file or directory exists:

```
>>> os.path.exists('memo.txt')
True
```

If it exists, `os.path.isdir` checks whether it's a directory:

```
>>> os.path.isdir('memo.txt')
False
>>> os.path.isdir('music')
True
```

Similarly, os.path.isfile checks whether it's a file.

os.listdir returns a list of the files (and other directories) in the given directory:

```
>>> os.listdir(cwd)
['music', 'photos', 'memo.txt']
```

To demonstrate these functions, the following example "walks" through a directory, prints the names of all the files, and calls itself recursively on all the directories.

```
def walk(dirname):
    for name in os.listdir(dirname):
        path = os.path.join(dirname, name)

        if os.path.isfile(path):
            print path
        else:
            walk(path)
```

os.path.join takes a directory and a file name and joins them into a complete path.

Exercise 14-1.

The os module provides a function called walk that is similar to this one but more versatile. Read the documentation and use it to print the names of the files in a given directory and its subdirectories.

Solution: *http://thinkpython.com/code/walk.py.*

Catching Exceptions

A lot of things can go wrong when you try to read and write files. If you try to open a file that doesn't exist, you get an IOError:

```
>>> fin = open('bad_file')
IOError: [Errno 2] No such file or directory: 'bad_file'
```

If you don't have permission to access a file:

```
>>> fout = open('/etc/passwd', 'w')
IOError: [Errno 13] Permission denied: '/etc/passwd'
```

And if you try to open a directory for reading, you get

```
>>> fin = open('/home')
IOError: [Errno 21] Is a directory
```

To avoid these errors, you could use functions like os.path.exists and os.path.is file, but it would take a lot of time and code to check all the possibilities (if "Errno 21" is any indication, there are at least 21 things that can go wrong).

It is better to go ahead and try—and deal with problems if they happen—which is exactly what the try statement does. The syntax is similar to an if statement:

```
try:
    fin = open('bad_file')
    for line in fin:
        print line
    fin.close()
except:
    print 'Something went wrong.'
```

Python starts by executing the try clause. If all goes well, it skips the except clause and proceeds. If an exception occurs, it jumps out of the try clause and executes the except clause.

Handling an exception with a try statement is called **catching** an exception. In this example, the except clause prints an error message that is not very helpful. In general, catching an exception gives you a chance to fix the problem, or try again, or at least end the program gracefully.

Exercise 14-2.
Write a function called sed that takes as arguments a pattern string, a replacement string, and two filenames; it should read the first file and write the contents into the second file (creating it if necessary). If the pattern string appears anywhere in the file, it should be replaced with the replacement string.

If an error occurs while opening, reading, writing or closing files, your program should catch the exception, print an error message, and exit. Solution: *http://thinkpython.com/ code/sed.py.*

Databases

A **database** is a file that is organized for storing data. Most databases are organized like a dictionary in the sense that they map from keys to values. The biggest difference is that the database is on disk (or other permanent storage), so it persists after the program ends.

The module anydbm provides an interface for creating and updating database files. As an example, I'll create a database that contains captions for image files.

Opening a database is similar to opening other files:

```
>>> import anydbm
>>> db = anydbm.open('captions.db', 'c')
```

The mode 'c' means that the database should be created if it doesn't already exist. The result is a database object that can be used (for most operations) like a dictionary. If you create a new item, anydbm updates the database file.

```
>>> db['cleese.png'] = 'Photo of John Cleese.'
```

When you access one of the items, anydbm reads the file:

```
>>> print db['cleese.png']
Photo of John Cleese.
```

If you make another assignment to an existing key, anydbm replaces the old value:

```
>>> db['cleese.png'] = 'Photo of John Cleese doing a silly walk.'
>>> print db['cleese.png']
Photo of John Cleese doing a silly walk.
```

Many dictionary methods, like keys and items, also work with database objects. So does iteration with a for statement.

```
for key in db:
    print key
```

As with other files, you should close the database when you are done:

```
>>> db.close()
```

Pickling

A limitation of anydbm is that the keys and values have to be strings. If you try to use any other type, you get an error.

The pickle module can help. It translates almost any type of object into a string suitable for storage in a database, and then translates strings back into objects.

pickle.dumps takes an object as a parameter and returns a string representation (dumps is short for "dump string"):

```
>>> import pickle
>>> t = [1, 2, 3]
>>> pickle.dumps(t)
'(lp0\nI1\naI2\naI3\na.'
```

The format isn't obvious to human readers; it is meant to be easy for pickle to interpret. pickle.loads ("load string") reconstitutes the object:

```
>>> t1 = [1, 2, 3]
>>> s = pickle.dumps(t1)
>>> t2 = pickle.loads(s)
>>> print t2
[1, 2, 3]
```

Although the new object has the same value as the old, it is not (in general) the same object:

```
>>> t1 == t2
True
>>> t1 is t2
False
```

In other words, pickling and then unpickling has the same effect as copying the object.

You can use pickle to store non-strings in a database. In fact, this combination is so common that it has been encapsulated in a module called shelve.

Exercise 14-3.
If you download my solution to Exercise 12-4 from *http://thinkpython.com/code/ anagram_sets.py*, you'll see that it creates a dictionary that maps from a sorted string of letters to the list of words that can be spelled with those letters. For example, 'opst' maps to the list ['opts', 'post', 'pots', 'spot', 'stop', 'tops'].

Write a module that imports anagram_sets and provides two new functions: store_an agrams should store the anagram dictionary in a "shelf;" read_anagrams should look up a word and return a list of its anagrams. Solution: *http://thinkpython.com/code/ anagram_db.py*

Pipes

Most operating systems provide a command-line interface, also known as a **shell**. Shells usually provide commands to navigate the file system and launch applications. For example, in Unix you can change directories with cd, display the contents of a directory with ls, and launch a web browser by typing (for example) firefox.

Any program that you can launch from the shell can also be launched from Python using a **pipe**. A pipe is an object that represents a running program.

For example, the Unix command ls -l normally displays the contents of the current directory (in long format). You can launch ls with os.popen[1]:

```
>>> cmd = 'ls -l'
>>> fp = os.popen(cmd)
```

The argument is a string that contains a shell command. The return value is an object that behaves just like an open file. You can read the output from the ls process one line at a time with readline or get the whole thing at once with read:

```
>>> res = fp.read()
```

When you are done, you close the pipe like a file:

```
>>> stat = fp.close()
>>> print stat
None
```

The return value is the final status of the ls process; None means that it ended normally (with no errors).

---

1. popen is deprecated now, which means we are supposed to stop using it and start using the subprocess module. But for simple cases, I find subprocess more complicated than necessary. So I am going to keep using popen until they take it away.

For example, most Unix systems provide a command called md5sum that reads the contents of a file and computes a "checksum." You can read about MD5 at *http://en.wikipedia.org/wiki/Md5*. This command provides an efficient way to check whether two files have the same contents. The probability that different contents yield the same checksum is very small (that is, unlikely to happen before the universe collapses).

You can use a pipe to run md5sum from Python and get the result:

```
>>> filename = 'book.tex'
>>> cmd = 'md5sum ' + filename
>>> fp = os.popen(cmd)
>>> res = fp.read()
>>> stat = fp.close()
>>> print res
1e0033f0ed0656636de0d75144ba32e0  book.tex
>>> print stat
None
```

Exercise 14-4.

In a large collection of MP3 files, there may be more than one copy of the same song, stored in different directories or with different file names. The goal of this exercise is to search for duplicates.

1. Write a program that searches a directory and all of its subdirectories, recursively, and returns a list of complete paths for all files with a given suffix (like .mp3). Hint: os.path provides several useful functions for manipulating file and path names.

2. To recognize duplicates, you can use md5sum to compute a "checksum" for each files. If two files have the same checksum, they probably have the same contents.

3. To double-check, you can use the Unix command diff.

Solution: *http://thinkpython.com/code/find_duplicates.py.*

Writing Modules

Any file that contains Python code can be imported as a module. For example, suppose you have a file named wc.py with the following code:

```
def linecount(filename):
    count = 0
    for line in open(filename):
        count += 1
    return count

print linecount('wc.py')
```

If you run this program, it reads itself and prints the number of lines in the file, which is 7. You can also import it like this:

```
>>> import wc
7
```

Now you have a module object wc:

```
>>> print wc
<module 'wc' from 'wc.py'>
```

That provides a function called linecount:

```
>>> wc.linecount('wc.py')
7
```

So that's how you write modules in Python.

The only problem with this example is that when you import the module it executes the test code at the bottom. Normally when you import a module, it defines new functions but it doesn't execute them.

Programs that will be imported as modules often use the following idiom:

```
if __name__ == '__main__':
    print linecount('wc.py')
```

__name__ is a built-in variable that is set when the program starts. If the program is running as a script, __name__ has the value __main__; in that case, the test code is executed. Otherwise, if the module is being imported, the test code is skipped.

Exercise 14-5.
Type this example into a file named wc.py and run it as a script. Then run the Python interpreter and import wc. What is the value of __name__ when the module is being imported?

Warning: If you import a module that has already been imported, Python does nothing. It does not re-read the file, even if it has changed.

If you want to reload a module, you can use the built-in function reload, but it can be tricky, so the safest thing to do is restart the interpreter and then import the module again.

Debugging

When you are reading and writing files, you might run into problems with whitespace. These errors can be hard to debug because spaces, tabs and newlines are normally invisible:

```
>>> s = '1 2\t 3\n 4'
>>> print s
1 2  3
 4
```

The built-in function `repr` can help. It takes any object as an argument and returns a string representation of the object. For strings, it represents whitespace characters with backslash sequences:

```
>>> print repr(s)
'1 2\t 3\n 4'
```

This can be helpful for debugging.

One other problem you might run into is that different systems use different characters to indicate the end of a line. Some systems use a newline, represented \n. Others use a return character, represented \r. Some use both. If you move files between different systems, these inconsistencies might cause problems.

For most systems, there are applications to convert from one format to another. You can find them (and read more about this issue) at *http://en.wikipedia.org/wiki/Newline*. Or, of course, you could write one yourself.

Glossary

Persistent:
> Pertaining to a program that runs indefinitely and keeps at least some of its data in permanent storage.

Format operator:
> An operator, %, that takes a format string and a tuple and generates a string that includes the elements of the tuple formatted as specified by the format string.

Format string:
> A string, used with the format operator, that contains format sequences.

Format sequence:
> A sequence of characters in a format string, like %d, that specifies how a value should be formatted.

Text file:
> A sequence of characters stored in permanent storage like a hard drive.

Directory:
> A named collection of files, also called a folder.

Path:
> A string that identifies a file.

Relative path:
> A path that starts from the current directory.

Absolute path:
> A path that starts from the topmost directory in the file system.

Catch:
   To prevent an exception from terminating a program using the `try` and `except` statements.

Database:
   A file whose contents are organized like a dictionary with keys that correspond to values.

Exercises

Exercise 14-6.
The `urllib` module provides methods for manipulating URLs and downloading information from the web. The following example downloads and prints a secret message from `thinkpython.com`:

```
import urllib

conn = urllib.urlopen('http://thinkpython.com/secret.html')
for line in conn:
    print line.strip()
```

Run this code and follow the instructions you see there. Solution: *http://thinkpython.com/code/zip_code.py.*

## CHAPTER 15

Classes and Objects

Code examples from this chapter are available from *http://thinkpython.com/code/ Point1.py*; solutions to the exercises are available from *http://thinkpython.com/code/ Point1_soln.py*.

User-Defined Types

We have used many of Python's built-in types; now we are going to define a new type. As an example, we will create a type called `Point` that represents a point in two-dimensional space.

In mathematical notation, points are often written in parentheses with a comma separating the coordinates. For example, *(0,0)* represents the origin, and *(x,y)* represents the point *x* units to the right and *y* units up from the origin.

There are several ways we might represent points in Python:

- We could store the coordinates separately in two variables, x and y.
- We could store the coordinates as elements in a list or tuple.
- We could create a new type to represent points as objects.

Creating a new type is (a little) more complicated than the other options, but it has advantages that will be apparent soon.

A user-defined type is also called a **class**. A class definition looks like this:

```
class Point(object):
    """Represents a point in 2-D space."""
```

This header indicates that the new class is a `Point`, which is a kind of `object`, which is a built-in type.

The body is a docstring that explains what the class is for. You can define variables and functions inside a class definition, but we will get back to that later.

Defining a class named `Point` creates a class object.

```
>>> print Point
<class '__main__.Point'>
```

Because `Point` is defined at the top level, its "full name" is `__main__.Point`.

The class object is like a factory for creating objects. To create a Point, you call `Point` as if it were a function.

```
>>> blank = Point()
>>> print blank
<__main__.Point instance at 0xb7e9d3ac>
```

The return value is a reference to a Point object, which we assign to `blank`. Creating a new object is called **instantiation**, and the object is an **instance** of the class.

When you print an instance, Python tells you what class it belongs to and where it is stored in memory (the prefix `0x` means that the following number is in hexadecimal).

Attributes

You can assign values to an instance using dot notation:

```
>>> blank.x = 3.0
>>> blank.y = 4.0
```

This syntax is similar to the syntax for selecting a variable from a module, such as `math.pi` or `string.whitespace`. In this case, though, we are assigning values to named elements of an object. These elements are called **attributes**.

As a noun, "AT-trib-ute" is pronounced with emphasis on the first syllable, as opposed to "a-TRIB-ute," which is a verb.

The following diagram shows the result of these assignments. A state diagram that shows an object and its attributes is called an **object diagram**; see Figure 15-1.

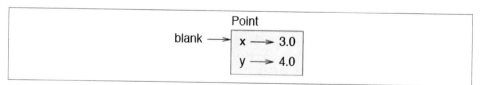

Figure 15-1. Object diagram.

The variable `blank` refers to a Point object, which contains two attributes. Each attribute refers to a floating-point number.

You can read the value of an attribute using the same syntax:

```
>>> print blank.y
4.0
>>> x = blank.x
>>> print x
3.0
```

The expression `blank.x` means, "Go to the object `blank` refers to and get the value of x." In this case, we assign that value to a variable named x. There is no conflict between the variable x and the attribute x.

You can use dot notation as part of any expression. For example:

```
>>> print '(%g, %g)' % (blank.x, blank.y)
(3.0, 4.0)
>>> distance = math.sqrt(blank.x**2 + blank.y**2)
>>> print distance
5.0
```

You can pass an instance as an argument in the usual way. For example:

```
def print_point(p):
    print '(%g, %g)' % (p.x, p.y)
```

`print_point` takes a point as an argument and displays it in mathematical notation. To invoke it, you can pass `blank` as an argument:

```
>>> print_point(blank)
(3.0, 4.0)
```

Inside the function, p is an alias for `blank`, so if the function modifies p, `blank` changes.

Exercise 15-1.
Write a function called `distance_between_points` that takes two Points as arguments and returns the distance between them.

Rectangles

Sometimes it is obvious what the attributes of an object should be, but other times you have to make decisions. For example, imagine you are designing a class to represent rectangles. What attributes would you use to specify the location and size of a rectangle? You can ignore angle; to keep things simple, assume that the rectangle is either vertical or horizontal.

There are at least two possibilities:

- You could specify one corner of the rectangle (or the center), the width, and the height.

- You could specify two opposing corners.

At this point it is hard to say whether either is better than the other, so we'll implement the first one, just as an example.

Here is the class definition:

```
class Rectangle(object):
    """Represents a rectangle.

    attributes: width, height, corner.
    """
```

The docstring lists the attributes: width and height are numbers; corner is a Point object that specifies the lower-left corner.

To represent a rectangle, you have to instantiate a Rectangle object and assign values to the attributes:

```
box = Rectangle()
box.width = 100.0
box.height = 200.0
box.corner = Point()
box.corner.x = 0.0
box.corner.y = 0.0
```

The expression box.corner.x means, "Go to the object box refers to and select the attribute named corner; then go to that object and select the attribute named x."

Figure 15-2 shows the state of this object. An object that is an attribute of another object is **embedded**.

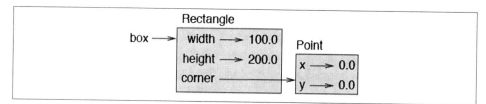

Figure 15-2. Object diagram.

Instances as Return Values

Functions can return instances. For example, find_center takes a Rectangle as an argument and returns a Point that contains the coordinates of the center of the Rectangle:

```
def find_center(rect):
    p = Point()
    p.x = rect.corner.x + rect.width/2.0
    p.y = rect.corner.y + rect.height/2.0
    return p
```

Here is an example that passes box as an argument and assigns the resulting Point to center:

```
>>> center = find_center(box)
>>> print_point(center)
(50.0, 100.0)
```

Objects Are Mutable

You can change the state of an object by making an assignment to one of its attributes. For example, to change the size of a rectangle without changing its position, you can modify the values of width and height:

```
box.width = box.width + 50
box.height = box.width + 100
```

You can also write functions that modify objects. For example, grow_rectangle takes a Rectangle object and two numbers, dwidth and dheight, and adds the numbers to the width and height of the rectangle:

```
def grow_rectangle(rect, dwidth, dheight):
    rect.width += dwidth
    rect.height += dheight
```

Here is an example that demonstrates the effect:

```
>>> print box.width
100.0
>>> print box.height
200.0
>>> grow_rectangle(box, 50, 100)
>>> print box.width
150.0
>>> print box.height
300.0
```

Inside the function, rect is an alias for box, so if the function modifies rect, box changes.

Exercise 15-2.
Write a function named move_rectangle that takes a Rectangle and two numbers named dx and dy. It should change the location of the rectangle by adding dx to the x coordinate of corner and adding dy to the y coordinate of corner.

Copying

Aliasing can make a program difficult to read because changes in one place might have unexpected effects in another place. It is hard to keep track of all the variables that might refer to a given object.

Copying an object is often an alternative to aliasing. The copy module contains a function called copy that can duplicate any object:

```
>>> p1 = Point()
>>> p1.x = 3.0
>>> p1.y = 4.0

>>> import copy
>>> p2 = copy.copy(p1)
```

p1 and p2 contain the same data, but they are not the same Point.

```
>>> print_point(p1)
(3.0, 4.0)
>>> print_point(p2)
(3.0, 4.0)
>>> p1 is p2
False
>>> p1 == p2
False
```

The is operator indicates that p1 and p2 are not the same object, which is what we expected. But you might have expected == to yield True because these points contain the same data. In that case, you will be disappointed to learn that for instances, the default behavior of the == operator is the same as the is operator; it checks object identity, not object equivalence. This behavior can be changed—we'll see how later.

If you use copy.copy to duplicate a Rectangle, you will find that it copies the Rectangle object but not the embedded Point.

```
>>> box2 = copy.copy(box)
>>> box2 is box
False
>>> box2.corner is box.corner
True
```

Figure 15-3 shows what the object diagram looks like. This operation is called a **shallow copy** because it copies the object and any references it contains, but not the embedded objects.

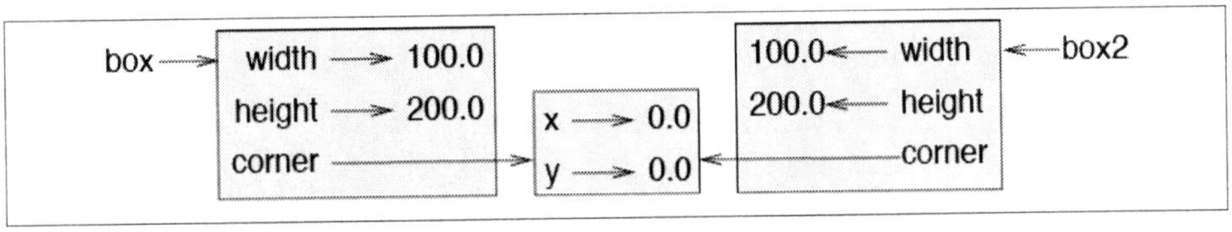

Figure 15-3. Object diagram.

For most applications, this is not what you want. In this example, invoking grow_rectangle on one of the Rectangles would not affect the other, but invoking move_rectangle on either would affect both! This behavior is confusing and error-prone.

Fortunately, the copy module contains a method named deepcopy that copies not only the object but also the objects it refers to, and the objects *they* refer to, and so on. You will not be surprised to learn that this operation is called a **deep copy**.

```
>>> box3 = copy.deepcopy(box)
>>> box3 is box
False
>>> box3.corner is box.corner
False
```

box3 and box are completely separate objects.

Exercise 15-3.
Write a version of move_rectangle that creates and returns a new Rectangle instead of modifying the old one.

Debugging

When you start working with objects, you are likely to encounter some new exceptions. If you try to access an attribute that doesn't exist, you get an AttributeError:

```
>>> p = Point()
>>> print p.z
AttributeError: Point instance has no attribute 'z'
```

If you are not sure what type an object is, you can ask:

```
>>> type(p)
<type '__main__.Point'>
```

If you are not sure whether an object has a particular attribute, you can use the built-in function hasattr:

```
>>> hasattr(p, 'x')
True
>>> hasattr(p, 'z')
False
```

The first argument can be any object; the second argument is a *string* that contains the name of the attribute.

Glossary

Class:
> A user-defined type. A class definition creates a new class object.

Class object:
> An object that contains information about a user-defined type. The class object can be used to create instances of the type.

Instance:
> An object that belongs to a class.

Attribute:
> One of the named values associated with an object.

Embedded (object):
> An object that is stored as an attribute of another object.

Shallow copy:
> To copy the contents of an object, including any references to embedded objects; implemented by the copy function in the copy module.

Deep copy:
> To copy the contents of an object as well as any embedded objects, and any objects embedded in them, and so on; implemented by the deepcopy function in the copy module.

Object diagram:
> A diagram that shows objects, their attributes, and the values of the attributes.

Exercises

Exercise 15-4.

Swampy (see Chapter 4) provides a module named World, which defines a user-defined type also called World. You can import it like this:

```
from swampy.World import World
```

The following code creates a World object and calls the mainloop method, which waits for the user.

```
world = World()
world.mainloop()
```

A window should appear with a title bar and an empty square. We will use this window to draw Points, Rectangles and other shapes. Add the following lines before calling `mainloop` and run the program again.

```
canvas = world.ca(width=500, height=500, background='white')
bbox = [[-150,-100], [150, 100]]
canvas.rectangle(bbox, outline='black', width=2, fill='green4')
```

You should see a green rectangle with a black outline. The first line creates a Canvas, which appears in the window as a white square. The Canvas object provides methods like `rectangle` for drawing various shapes.

`bbox` is a list of lists that represents the "bounding box" of the rectangle. The first pair of coordinates is the lower-left corner of the rectangle; the second pair is the upper-right corner.

You can draw a circle like this:

```
canvas.circle([-25,0], 70, outline=None, fill='red')
```

The first parameter is the coordinate pair for the center of the circle; the second parameter is the radius.

If you add this line to the program, the result should resemble the national flag of Bangladesh (see *http://en.wikipedia.org/wiki/Gallery_of_sovereign-state_flags*).

1. Write a function called `draw_rectangle` that takes a Canvas and a Rectangle as arguments and draws a representation of the Rectangle on the Canvas.

2. Add an attribute named `color` to your Rectangle objects and modify `draw_rectangle` so that it uses the color attribute as the fill color.

3. Write a function called `draw_point` that takes a Canvas and a Point as arguments and draws a representation of the Point on the Canvas.

4. Define a new class called Circle with appropriate attributes and instantiate a few Circle objects. Write a function called `draw_circle` that draws circles on the canvas.

5. Write a program that draws the national flag of the Czech Republic. Hint: you can draw a polygon like this:

```
points = [[-150,-100], [150, 100], [150, -100]]
canvas.polygon(points, fill='blue')
```

I have written a small program that lists the available colors; you can download it from *http://thinkpython.com/code/color_list.py*.

Classes and Functions

Code examples from this chapter are available from *http://thinkpython.com/code/ Time1.py.*

Time

As another example of a user-defined type, we'll define a class called Time that records the time of day. The class definition looks like this:

```
class Time(object):
    """Represents the time of day.

    attributes: hour, minute, second
    """
```

We can create a new Time object and assign attributes for hours, minutes, and seconds:

```
time = Time()
time.hour = 11
time.minute = 59
time.second = 30
```

The state diagram for the Time object looks like Figure 16-1.

Exercise 16-1.
Write a function called print_time that takes a Time object and prints it in the form hour:minute:second. Hint: the format sequence '%.2d' prints an integer using at least two digits, including a leading zero if necessary.

Exercise 16-2.
Write a boolean function called is_after that takes two Time objects, t1 and t2, and returns True if t1 follows t2 chronologically and False otherwise. Challenge: don't use an if statement.

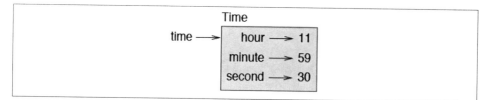

Figure 16-1. Object diagram.

Pure Functions

In the next few sections, we'll write two functions that add time values. They demonstrate two kinds of functions: pure functions and modifiers. They also demonstrate a development plan I'll call **prototype and patch**, which is a way of tackling a complex problem by starting with a simple prototype and incrementally dealing with the complications.

Here is a simple prototype of add_time:

```
def add_time(t1, t2):
    sum = Time()
    sum.hour = t1.hour + t2.hour
    sum.minute = t1.minute + t2.minute
    sum.second = t1.second + t2.second
    return sum
```

The function creates a new Time object, initializes its attributes, and returns a reference to the new object. This is called a **pure function** because it does not modify any of the objects passed to it as arguments and it has no effect, like displaying a value or getting user input, other than returning a value.

To test this function, I'll create two Time objects: start contains the start time of a movie, like *Monty Python and the Holy Grail*, and duration contains the run time of the movie, which is one hour 35 minutes.

add_time figures out when the movie will be done.

```
>>> start = Time()
>>> start.hour = 9
>>> start.minute = 45
>>> start.second =  0

>>> duration = Time()
>>> duration.hour = 1
>>> duration.minute = 35
>>> duration.second = 0

>>> done = add_time(start, duration)
>>> print_time(done)
10:80:00
```

The result, 10:80:00 might not be what you were hoping for. The problem is that this function does not deal with cases where the number of seconds or minutes adds up to more than sixty. When that happens, we have to "carry" the extra seconds into the minute column or the extra minutes into the hour column.

Here's an improved version:

```
def add_time(t1, t2):
    sum = Time()
    sum.hour = t1.hour + t2.hour
    sum.minute = t1.minute + t2.minute
    sum.second = t1.second + t2.second

    if sum.second >= 60:
        sum.second -= 60
        sum.minute += 1

    if sum.minute >= 60:
        sum.minute -= 60
        sum.hour += 1

    return sum
```

Although this function is correct, it is starting to get big. We will see a shorter alternative later.

Modifiers

Sometimes it is useful for a function to modify the objects it gets as parameters. In that case, the changes are visible to the caller. Functions that work this way are called **modifiers**.

increment, which adds a given number of seconds to a Time object, can be written naturally as a modifier. Here is a rough draft:

```
def increment(time, seconds):
    time.second += seconds

    if time.second >= 60:
        time.second -= 60
        time.minute += 1

    if time.minute >= 60:
        time.minute -= 60
        time.hour += 1
```

The first line performs the basic operation; the remainder deals with the special cases we saw before.

Is this function correct? What happens if the parameter seconds is much greater than sixty?

In that case, it is not enough to carry once; we have to keep doing it until `time.sec` `ond` is less than sixty. One solution is to replace the `if` statements with `while` statements. That would make the function correct, but not very efficient.

Exercise 16-3.
Write a correct version of `increment` that doesn't contain any loops.

Anything that can be done with modifiers can also be done with pure functions. In fact, some programming languages only allow pure functions. There is some evidence that programs that use pure functions are faster to develop and less error-prone than programs that use modifiers. But modifiers are convenient at times, and functional programs tend to be less efficient.

In general, I recommend that you write pure functions whenever it is reasonable and resort to modifiers only if there is a compelling advantage. This approach might be called a **functional programming style**.

Exercise 16-4.
Write a "pure" version of `increment` that creates and returns a new Time object rather than modifying the parameter.

Prototyping Versus Planning

The development plan I am demonstrating is called "prototype and patch." For each function, I wrote a prototype that performed the basic calculation and then tested it, patching errors along the way.

This approach can be effective, especially if you don't yet have a deep understanding of the problem. But incremental corrections can generate code that is unnecessarily complicated—since it deals with many special cases—and unreliable—since it is hard to know if you have found all the errors.

An alternative is **planned development**, in which high-level insight into the problem can make the programming much easier. In this case, the insight is that a Time object is really a three-digit number in base 60 (see *http://en.wikipedia.org/wiki/Sexagesimal.*)! The `second` attribute is the "ones column," the `minute` attribute is the "sixties column," and the hour attribute is the "thirty-six hundreds column."

When we wrote `add_time` and `increment`, we were effectively doing addition in base 60, which is why we had to carry from one column to the next.

This observation suggests another approach to the whole problem—we can convert Time objects to integers and take advantage of the fact that the computer knows how to do integer arithmetic.

Here is a function that converts Times to integers:

```
def time_to_int(time):
    minutes = time.hour * 60 + time.minute
    seconds = minutes * 60 + time.second
    return seconds
```

And here is the function that converts integers to Times (recall that `divmod` divides the first argument by the second and returns the quotient and remainder as a tuple).

```
def int_to_time(seconds):
    time = Time()
    minutes, time.second = divmod(seconds, 60)
    time.hour, time.minute = divmod(minutes, 60)
    return time
```

You might have to think a bit, and run some tests, to convince yourself that these functions are correct. One way to test them is to check that `time_to_int(int_to_time(x))` `== x` for many values of x. This is an example of a consistency check.

Once you are convinced they are correct, you can use them to rewrite `add_time`:

```
def add_time(t1, t2):
    seconds = time_to_int(t1) + time_to_int(t2)
    return int_to_time(seconds)
```

This version is shorter than the original, and easier to verify.

Exercise 16-5.
Rewrite `increment` using `time_to_int` and `int_to_time`.

In some ways, converting from base 60 to base 10 and back is harder than just dealing with times. Base conversion is more abstract; our intuition for dealing with time values is better.

But if we have the insight to treat times as base 60 numbers and make the investment of writing the conversion functions (`time_to_int` and `int_to_time`), we get a program that is shorter, easier to read and debug, and more reliable.

It is also easier to add features later. For example, imagine subtracting two Times to find the duration between them. The naive approach would be to implement subtraction with borrowing. Using the conversion functions would be easier and more likely to be correct.

Ironically, sometimes making a problem harder (or more general) makes it easier (because there are fewer special cases and fewer opportunities for error).

Debugging

A Time object is well-formed if the values of `minute` and `second` are between 0 and 60 (including 0 but not 60) and if `hour` is positive. `hour` and `minute` should be integral values, but we might allow `second` to have a fraction part.

Requirements like these are called **invariants** because they should always be true. To put it a different way, if they are not true, then something has gone wrong.

Writing code to check your invariants can help you detect errors and find their causes. For example, you might have a function like `valid_time` that takes a Time object and returns `False` if it violates an invariant:

```
def valid_time(time):
    if time.hour < 0 or time.minute < 0 or time.second < 0:
        return False
    if time.minute >= 60 or time.second >= 60:
        return False
    return True
```

Then at the beginning of each function you could check the arguments to make sure they are valid:

```
def add_time(t1, t2):
    if not valid_time(t1) or not valid_time(t2):
        raise ValueError, 'invalid Time object in add_time'
    seconds = time_to_int(t1) + time_to_int(t2)
    return int_to_time(seconds)
```

Or you could use an `assert` statement, which checks a given invariant and raises an exception if it fails:

```
def add_time(t1, t2):
    assert valid_time(t1) and valid_time(t2)
    seconds = time_to_int(t1) + time_to_int(t2)
    return int_to_time(seconds)
```

`assert` statements are useful because they distinguish code that deals with normal conditions from code that checks for errors.

Glossary

Prototype and patch:
  A development plan that involves writing a rough draft of a program, testing, and correcting errors as they are found.

Planned development:
  A development plan that involves high-level insight into the problem and more planning than incremental development or prototype development.

Pure function:
  A function that does not modify any of the objects it receives as arguments. Most pure functions are fruitful.

Modifier:
  A function that changes one or more of the objects it receives as arguments. Most modifiers are fruitless.

Functional programming style:
A style of program design in which the majority of functions are pure.

Invariant:
A condition that should always be true during the execution of a program.

Exercises

Code examples from this chapter are available from *http://thinkpython.com/code/ Time1.py*; solutions to these exercises are available from *http://thinkpython.com/code/ Time1_soln.py*.

Exercise 16-6.
Write a function called `mul_time` that takes a Time object and a number and returns a new Time object that contains the product of the original Time and the number.

Then use `mul_time` to write a function that takes a Time object that represents the finishing time in a race, and a number that represents the distance, and returns a Time object that represents the average pace (time per mile).

Exercise 16-7.
The `datetime` module provides `date` and `time` objects that are similar to the Date and Time objects in this chapter, but they provide a rich set of methods and operators. Read the documentation at *http://docs.python.org/lib/datetime-date.html*.

1. Use the `datetime` module to write a program that gets the current date and prints the day of the week.

2. Write a program that takes a birthday as input and prints the user's age and the number of days, hours, minutes and seconds until their next birthday.

3. For two people born on different days, there is a day when one is twice as old as the other. That's their Double Day. Write a program that takes two birthdays and computes their Double Day.

4. For a little more challenge, write the more general version that computes the day when one person is *n* times older than the other.

CHAPTER 17

Classes and Methods

Code examples from this chapter are available from *http://thinkpython.com/code/Time2.py.*

Object-Oriented Features

Python is an **object-oriented programming language**, which means that it provides features that support object-oriented programming.

It is not easy to define object-oriented programming, but we have already seen some of its characteristics:

- Programs are made up of object definitions and function definitions, and most of the computation is expressed in terms of operations on objects.
- Each object definition corresponds to some object or concept in the real world, and the functions that operate on that object correspond to the ways real-world objects interact.

For example, the Time class defined in Chapter 16 corresponds to the way people record the time of day, and the functions we defined correspond to the kinds of things people do with times. Similarly, the Point and Rectangle classes correspond to the mathematical concepts of a point and a rectangle.

So far, we have not taken advantage of the features Python provides to support object-oriented programming. These features are not strictly necessary; most of them provide alternative syntax for things we have already done. But in many cases, the alternative is more concise and more accurately conveys the structure of the program.

For example, in the Time program, there is no obvious connection between the class definition and the function definitions that follow. With some examination, it is apparent that every function takes at least one Time object as an argument.

This observation is the motivation for **methods**; a method is a function that is associated with a particular class. We have seen methods for strings, lists, dictionaries and tuples. In this chapter, we will define methods for user-defined types.

Methods are semantically the same as functions, but there are two syntactic differences:

- Methods are defined inside a class definition in order to make the relationship between the class and the method explicit.
- The syntax for invoking a method is different from the syntax for calling a function.

In the next few sections, we will take the functions from the previous two chapters and transform them into methods. This transformation is purely mechanical; you can do it simply by following a sequence of steps. If you are comfortable converting from one form to another, you will be able to choose the best form for whatever you are doing.

Printing Objects

In Chapter 16, we defined a class named Time and in Exercise 16-1, you wrote a function named print_time:

```
class Time(object):
    """Represents the time of day."""

def print_time(time):
    print '%.2d:%.2d:%.2d' % (time.hour, time.minute, time.second)
```

To call this function, you have to pass a Time object as an argument:

```
>>> start = Time()
>>> start.hour = 9
>>> start.minute = 45
>>> start.second = 00
>>> print_time(start)
09:45:00
```

To make print_time a method, all we have to do is move the function definition inside the class definition. Notice the change in indentation.

```
class Time(object):
    def print_time(time):
        print '%.2d:%.2d:%.2d' % (time.hour, time.minute, time.second)
```

Now there are two ways to call print_time. The first (and less common) way is to use function syntax:

```
>>> Time.print_time(start)
09:45:00
```

In this use of dot notation, Time is the name of the class, and print_time is the name of the method. start is passed as a parameter.

The second (and more concise) way is to use method syntax:

```
>>> start.print_time()
09:45:00
```

In this use of dot notation, `print_time` is the name of the method (again), and `start` is the object the method is invoked on, which is called the **subject**. Just as the subject of a sentence is what the sentence is about, the subject of a method invocation is what the method is about.

Inside the method, the subject is assigned to the first parameter, so in this case `start` is assigned to `time`.

By convention, the first parameter of a method is called `self`, so it would be more common to write `print_time` like this:

```
class Time(object):
    def print_time(self):
        print '%.2d:%.2d:%.2d' % (self.hour, self.minute, self.second)
```

The reason for this convention is an implicit metaphor:

- The syntax for a function call, `print_time(start)`, suggests that the function is the active agent. It says something like, "Hey `print_time`! Here's an object for you to print."
- In object-oriented programming, the objects are the active agents. A method invocation like `start.print_time()` says "Hey `start`! Please print yourself."

This change in perspective might be more polite, but it is not obvious that it is useful. In the examples we have seen so far, it may not be. But sometimes shifting responsibility from the functions onto the objects makes it possible to write more versatile functions, and makes it easier to maintain and reuse code.

Exercise 17-1.
Rewrite `time_to_int` (from "Prototyping Versus Planning" (page 184)) as a method. It is probably not appropriate to rewrite `int_to_time` as a method; what object you would invoke it on?

Another Example

Here's a version of `increment` (from "Modifiers" (page 183)) rewritten as a method:

```
# inside class Time:

    def increment(self, seconds):
        seconds += self.time_to_int()
        return int_to_time(seconds)
```

This version assumes that `time_to_int` is written as a method, as in Exercise 17-1. Also, note that it is a pure function, not a modifier.

Here's how you would invoke `increment`:

```
>>> start.print_time()
09:45:00
>>> end = start.increment(1337)
>>> end.print_time()
10:07:17
```

The subject, `start`, gets assigned to the first parameter, `self`. The argument, 1337, gets assigned to the second parameter, `seconds`.

This mechanism can be confusing, especially if you make an error. For example, if you invoke `increment` with two arguments, you get:

```
>>> end = start.increment(1337, 460)
TypeError: increment() takes exactly 2 arguments (3 given)
```

The error message is initially confusing, because there are only two arguments in parentheses. But the subject is also considered an argument, so all together that's three.

A More Complicated Example

`is_after` (from Exercise 16-2) is slightly more complicated because it takes two Time objects as parameters. In this case it is conventional to name the first parameter `self` and the second parameter `other`:

```
# inside class Time:

    def is_after(self, other):
        return self.time_to_int() > other.time_to_int()
```

To use this method, you have to invoke it on one object and pass the other as an argument:

```
>>> end.is_after(start)
True
```

One nice thing about this syntax is that it almost reads like English: "end is after start?"

The init Method

The init method (short for "initialization") is a special method that gets invoked when an object is instantiated. Its full name is __init__ (two underscore characters, followed by init, and then two more underscores). An init method for the Time class might look like this:

```
# inside class Time:

    def __init__(self, hour=0, minute=0, second=0):
        self.hour = hour
        self.minute = minute
        self.second = second
```

It is common for the parameters of __init__ to have the same names as the attributes. The statement

```
        self.hour = hour
```

stores the value of the parameter hour as an attribute of self.

The parameters are optional, so if you call Time with no arguments, you get the default values.

```
>>> time = Time()
>>> time.print_time()
00:00:00
```

If you provide one argument, it overrides hour:

```
>>> time = Time (9)
>>> time.print_time()
09:00:00
```

If you provide two arguments, they override hour and minute.

```
>>> time = Time(9, 45)
>>> time.print_time()
09:45:00
```

And if you provide three arguments, they override all three default values.

Exercise 17-2.
Write an init method for the Point class that takes x and y as optional parameters and assigns them to the corresponding attributes.

The __str__ Method

__str__ is a special method, like __init__, that is supposed to return a string representation of an object.

For example, here is a str method for Time objects:

```
# inside class Time:

    def __str__(self):
        return '%.2d:%.2d:%.2d' % (self.hour, self.minute, self.second)
```

When you print an object, Python invokes the str method:

```
>>> time = Time(9, 45)
>>> print time
09:45:00
```

When I write a new class, I almost always start by writing __init__, which makes it easier to instantiate objects, and __str__, which is useful for debugging.

Exercise 17-3.
Write a str method for the Point class. Create a Point object and print it.

Operator Overloading

By defining other special methods, you can specify the behavior of operators on user-defined types. For example, if you define a method named __add__ for the Time class, you can use the + operator on Time objects.

Here is what the definition might look like:

```
# inside class Time:

    def __add__(self, other):
        seconds = self.time_to_int() + other.time_to_int()
        return int_to_time(seconds)
```

And here is how you could use it:

```
>>> start = Time(9, 45)
>>> duration = Time(1, 35)
>>> print start + duration
11:20:00
```

When you apply the + operator to Time objects, Python invokes __add__. When you print the result, Python invokes __str__. So there is quite a lot happening behind the scenes!

Changing the behavior of an operator so that it works with user-defined types is called **operator overloading**. For every operator in Python there is a corresponding special method, like __add__. For more details, see *http://docs.python.org/ref/specialnames.html.*

Exercise 17-4.
Write an add method for the Point class.

Type-Based Dispatch

In the previous section we added two Time objects, but you also might want to add an integer to a Time object. The following is a version of __add__ that checks the type of other and invokes either add_time or increment:

```
# inside class Time:

    def __add__(self, other):
        if isinstance(other, Time):
            return self.add_time(other)
        else:
            return self.increment(other)

    def add_time(self, other):
        seconds = self.time_to_int() + other.time_to_int()
        return int_to_time(seconds)

    def increment(self, seconds):
        seconds += self.time_to_int()
        return int_to_time(seconds)
```

The built-in function isinstance takes a value and a class object, and returns True if the value is an instance of the class.

If other is a Time object, __add__ invokes add_time. Otherwise it assumes that the parameter is a number and invokes increment. This operation is called a **type-based dispatch** because it dispatches the computation to different methods based on the type of the arguments.

Here are examples that use the + operator with different types:

```
>>> start = Time(9, 45)
>>> duration = Time(1, 35)
>>> print start + duration
11:20:00
>>> print start + 1337
10:07:17
```

Unfortunately, this implementation of addition is not commutative. If the integer is the first operand, you get

```
>>> print 1337 + start
TypeError: unsupported operand type(s) for +: 'int' and 'instance'
```

The problem is, instead of asking the Time object to add an integer, Python is asking an integer to add a Time object, and it doesn't know how to do that. But there is a clever solution for this problem: the special method __radd__, which stands for "right-side add." This method is invoked when a Time object appears on the right side of the + operator. Here's the definition:

```
# inside class Time:

    def __radd__(self, other):
        return self.__add__(other)
```

And here's how it's used:

```
>>> print 1337 + start
10:07:17
```

Exercise 17-5.

Write an add method for Points that works with either a Point object or a tuple:

- If the second operand is a Point, the method should return a new Point whose x coordinate is the sum of the x coordinates of the operands, and likewise for the y coordinates.

- If the second operand is a tuple, the method should add the first element of the tuple to the x coordinate and the second element to the y coordinate, and return a new Point with the result.

Polymorphism

Type-based dispatch is useful when it is necessary, but (fortunately) it is not always necessary. Often you can avoid it by writing functions that work correctly for arguments with different types.

Many of the functions we wrote for strings will actually work for any kind of sequence. For example, in "Dictionary as a Set of Counters" (page 123) we used histogram to count the number of times each letter appears in a word.

```
def histogram(s):
    d = dict()
    for c in s:
        if c not in d:
            d[c] = 1
        else:
            d[c] = d[c]+1
    return d
```

This function also works for lists, tuples, and even dictionaries, as long as the elements of s are hashable, so they can be used as keys in d.

```
>>> t = ['spam', 'egg', 'spam', 'spam', 'bacon', 'spam']
>>> histogram(t)
{'bacon': 1, 'egg': 1, 'spam': 4}
```

Functions that can work with several types are called **polymorphic**. Polymorphism can facilitate code reuse. For example, the built-in function sum, which adds the elements of a sequence, works as long as the elements of the sequence support addition.

Since Time objects provide an add method, they work with sum:

```
>>> t1 = Time(7, 43)
>>> t2 = Time(7, 41)
```

```
>>> t3 = Time(7, 37)
>>> total = sum([t1, t2, t3])
>>> print total
23:01:00
```

In general, if all of the operations inside a function work with a given type, then the function works with that type.

The best kind of polymorphism is the unintentional kind, where you discover that a function you already wrote can be applied to a type you never planned for.

Debugging

It is legal to add attributes to objects at any point in the execution of a program, but if you are a stickler for type theory, it is a dubious practice to have objects of the same type with different attribute sets. It is usually a good idea to initialize all of an objects attributes in the init method.

If you are not sure whether an object has a particular attribute, you can use the built-in function hasattr (see "Debugging" (page 177)).

Another way to access the attributes of an object is through the special attribute __dict__, which is a dictionary that maps attribute names (as strings) and values:

```
>>> p = Point(3, 4)
>>> print p.__dict__
{'y': 4, 'x': 3}
```

For purposes of debugging, you might find it useful to keep this function handy:

```
def print_attributes(obj):
    for attr in obj.__dict__:
        print attr, getattr(obj, attr)
```

print_attributes traverses the items in the object's dictionary and prints each attribute name and its corresponding value.

The built-in function getattr takes an object and an attribute name (as a string) and returns the attribute's value.

Interface and Implementation

One of the goals of object-oriented design is to make software more maintainable, which means that you can keep the program working when other parts of the system change, and modify the program to meet new requirements.

A design principle that helps achieve that goal is to keep interfaces separate from implementations. For objects, that means that the methods a class provides should not depend on how the attributes are represented.

For example, in this chapter we developed a class that represents a time of day. Methods provided by this class include and `time_to_int`, `is_after`, and `add_time`.

We could implement those methods in several ways. The details of the implementation depend on how we represent time. In this chapter, the attributes of a `Time` object are `hour`, `minute`, and `second`.

As an alternative, we could replace these attributes with a single integer representing the number of seconds since midnight. This implementation would make some methods, like `is_after`, easier to write, but it makes some methods harder.

After you deploy a new class, you might discover a better implementation. If other parts of the program are using your class, it might be time-consuming and error-prone to change the interface.

But if you designed the interface carefully, you can change the implementation without changing the interface, which means that other parts of the program don't have to change.

Keeping the interface separate from the implementation means that you have to hide the attributes. Code in other parts of the program (outside the class definition) should use methods to read and modify the state of the object. They should not access the attributes directly. This principle is called **information hiding**; see *http:// en.wikipedia.org/wiki/Information_hiding*.

Exercise 17-6.

Download the code from this chapter (*http://thinkpython.com/code/Time2.py*). Change the attributes of `Time` to be a single integer representing seconds since midnight. Then modify the methods (and the function `int_to_time`) to work with the new implementation. You should not have to modify the test code in `main`. When you are done, the output should be the same as before. Solution: *http://thinkpython.com/code/Time2_soln.py*

Glossary

Object-oriented language:
   A language that provides features, such as user-defined classes and method syntax, that facilitate object-oriented programming.

Object-oriented programming:
   A style of programming in which data and the operations that manipulate it are organized into classes and methods.

Method:
   A function that is defined inside a class definition and is invoked on instances of that class.

Subject:
The object a method is invoked on.

Operator overloading:
Changing the behavior of an operator like + so it works with a user-defined type.

Type-based dispatch:
A programming pattern that checks the type of an operand and invokes different functions for different types.

Polymorphic:
Pertaining to a function that can work with more than one type.

Information hiding:
The principle that the interface provided by an object should not depend on its implementation, in particular the representation of its attributes.

Exercises

Exercise 17-7.
This exercise is a cautionary tale about one of the most common, and difficult to find, errors in Python.

Write a definition for a class named Kangaroo with the following methods:

1. An __init__ method that initializes an attribute named pouch_contents to an empty list.

2. A method named put_in_pouch that takes an object of any type and adds it to pouch_contents.

3. A __str__ method that returns a string representation of the Kangaroo object and the contents of the pouch.

Test your code by creating two Kangaroo objects, assigning them to variables named kanga and roo, and then adding roo to the contents of kanga's pouch.

Download *http://thinkpython.com/code/BadKangaroo.py*. It contains a solution to the previous problem with one big, nasty bug. Find and fix the bug.

If you get stuck, you can download *http://thinkpython.com/code/GoodKangaroo.py*, which explains the problem and demonstrates a solution.

Exercise 17-8.
Visual is a Python module that provides 3-D graphics. It is not always included in a Python installation, so you might have to install it from your software repository or, if it's not there, from *http://vpython.org*.

The following example creates a 3-D space that is 256 units wide, long and high, and sets the "center" to be the point *(128,128,128)*. Then it draws a blue sphere.

```
from visual import *

scene.range = (256, 256, 256)
scene.center = (128, 128, 128)

color = (0.1, 0.1, 0.9)          # mostly blue
sphere(pos=scene.center, radius=128, color=color)
```

color is an RGB tuple; that is, the elements are Red-Green-Blue levels between 0.0 and 1.0 (see *http://en.wikipedia.org/wiki/RGB_color_model*).

If you run this code, you should see a window with a black background and a blue sphere. If you drag the middle button up and down, you can zoom in and out. You can also rotate the scene by dragging the right button, but with only one sphere in the world, it is hard to tell the difference.

The following loop creates a cube of spheres:

```
t = range(0, 256, 51)
for x in t:
    for y in t:
        for z in t:
            pos = x, y, z
            sphere(pos=pos, radius=10, color=color)
```

1. Put this code in a script and make sure it works for you.

2. Modify the program so that each sphere in the cube has the color that corresponds to its position in RGB space. Notice that the coordinates are in the range 0–255, but the RGB tuples are in the range 0.0–1.0.

3. Download *http://thinkpython.com/code/color_list.py* and use the function read_colors to generate a list of the available colors on your system, their names and RGB values. For each named color draw a sphere in the position that corresponds to its RGB values.

You can see my solution at *http://thinkpython.com/code/color_space.py*.

## CHAPTER 18

Inheritance

In this chapter I present classes to represent playing cards, decks of cards, and poker hands. If you don't play poker, you can read about it at *http://en.wikipedia.org/wiki/Poker*, but you don't have to; I'll tell you what you need to know for the exercises. Code examples from this chapter are available from *http://thinkpython.com/code/Card.py*.

If you are not familiar with Anglo-American playing cards, you can read about them at *http://en.wikipedia.org/wiki/Playing_cards*.

Card Objects

There are fifty-two cards in a deck, each of which belongs to one of four suits and one of thirteen ranks. The suits are Spades, Hearts, Diamonds, and Clubs (in descending order in bridge). The ranks are Ace, 2, 3, 4, 5, 6, 7, 8, 9, 10, Jack, Queen, and King. Depending on the game that you are playing, an Ace may be higher than King or lower than 2.

If we want to define a new object to represent a playing card, it is obvious what the attributes should be: rank and suit. It is not as obvious what type the attributes should be. One possibility is to use strings containing words like 'Spade' for suits and 'Queen' for ranks. One problem with this implementation is that it would not be easy to compare cards to see which had a higher rank or suit.

An alternative is to use integers to **encode** the ranks and suits. In this context, "encode" means that we are going to define a mapping between numbers and suits, or between numbers and ranks. This kind of encoding is not meant to be a secret (that would be "encryption").

For example, this table shows the suits and the corresponding integer codes:

Spades   ⟼   3
Hearts   ⟼   2
Diamonds ⟼   1
Clubs   ⟼   0

This code makes it easy to compare cards; because higher suits map to higher numbers, we can compare suits by comparing their codes.

The mapping for ranks is fairly obvious; each of the numerical ranks maps to the corresponding integer, and for face cards:

Jack   ⟼   11
Queen ⟼   12
King   ⟼   13

I am using the ⟼ symbol to make it clear that these mappings are not part of the Python program. They are part of the program design, but they don't appear explicitly in the code.

The class definition for Card looks like this:

```
class Card(object):
    """Represents a standard playing card."""

    def __init__(self, suit=0, rank=2):
        self.suit = suit
        self.rank = rank
```

As usual, the init method takes an optional parameter for each attribute. The default card is the 2 of Clubs.

To create a Card, you call Card with the suit and rank of the card you want.

```
queen_of_diamonds = Card(1, 12)
```

Class Attributes

In order to print Card objects in a way that people can easily read, we need a mapping from the integer codes to the corresponding ranks and suits. A natural way to do that is with lists of strings. We assign these lists to **class attributes**:

```
# inside class Card:

    suit_names = ['Clubs', 'Diamonds', 'Hearts', 'Spades']
    rank_names = [None, 'Ace', '2', '3', '4', '5', '6', '7',
              '8', '9', '10', 'Jack', 'Queen', 'King']

    def __str__(self):
        return '%s of %s' % (Card.rank_names[self.rank],
                             Card.suit_names[self.suit])
```

Variables like `suit_names` and `rank_names`, which are defined inside a class but outside of any method, are called class attributes because they are associated with the class object `Card`.

This term distinguishes them from variables like `suit` and `rank`, which are called **instance attributes** because they are associated with a particular instance.

Both kinds of attribute are accessed using dot notation. For example, in `__str__`, `self` is a Card object, and `self.rank` is its rank. Similarly, `Card` is a class object, and `Card.rank_names` is a list of strings associated with the class.

Every card has its own `suit` and `rank`, but there is only one copy of `suit_names` and `rank_names`.

Putting it all together, the expression `Card.rank_names[self.rank]` means "use the attribute `rank` from the object `self` as an index into the list `rank_names` from the class `Card`, and select the appropriate string."

The first element of `rank_names` is `None` because there is no card with rank zero. By including `None` as a place-keeper, we get a mapping with the nice property that the index 2 maps to the string `'2'`, and so on. To avoid this tweak, we could have used a dictionary instead of a list.

With the methods we have so far, we can create and print cards:

```
>>> card1 = Card(2, 11)
>>> print card1
Jack of Hearts
```

Figure 18-1 is a diagram of the `Card` class object and one Card instance. `Card` is a class object, so it has type `type`. `card1` has type `Card`. (To save space, I didn't draw the contents of `suit_names` and `rank_names`).

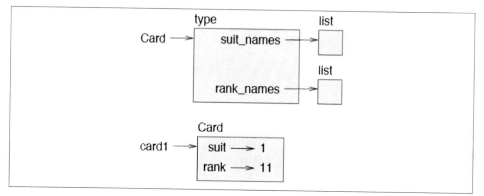

Figure 18-1. Object diagram.

Comparing Cards

For built-in types, there are relational operators (<, >, ==, etc.) that compare values and determine when one is greater than, less than, or equal to another. For user-defined types, we can override the behavior of the built-in operators by providing a method named __cmp__.

__cmp__ takes two parameters, self and other, and returns a positive number if the first object is greater, a negative number if the second object is greater, and 0 if they are equal to each other.

The correct ordering for cards is not obvious. For example, which is better, the 3 of Clubs or the 2 of Diamonds? One has a higher rank, but the other has a higher suit. In order to compare cards, you have to decide whether rank or suit is more important.

The answer might depend on what game you are playing, but to keep things simple, we'll make the arbitrary choice that suit is more important, so all of the Spades outrank all of the Diamonds, and so on.

With that decided, we can write __cmp__:

```
# inside class Card:

    def __cmp__(self, other):
        # check the suits
        if self.suit > other.suit: return 1
        if self.suit < other.suit: return -1

        # suits are the same... check ranks
        if self.rank > other.rank: return 1
        if self.rank < other.rank: return -1

        # ranks are the same... it's a tie
        return 0
```

You can write this more concisely using tuple comparison:

```
# inside class Card:

def __cmp__(self, other):
    t1 = self.suit, self.rank
    t2 = other.suit, other.rank
    return cmp(t1, t2)
```

The built-in function cmp has the same interface as the method __cmp__: it takes two values and returns a positive number if the first is larger, a negative number if the second is larger, and 0 if they are equal.

Exercise 18-1.
Write a __cmp__ method for Time objects. Hint: you can use tuple comparison, but you also might consider using integer subtraction.

Decks

Now that we have Cards, the next step is to define Decks. Since a deck is made up of cards, it is natural for each Deck to contain a list of cards as an attribute.

The following is a class definition for Deck. The init method creates the attribute cards and generates the standard set of fifty-two cards:

```
class Deck(object):

    def __init__(self):
        self.cards = []
        for suit in range(4):
            for rank in range(1, 14):
                card = Card(suit, rank)
                self.cards.append(card)
```

The easiest way to populate the deck is with a nested loop. The outer loop enumerates the suits from 0 to 3. The inner loop enumerates the ranks from 1 to 13. Each iteration creates a new Card with the current suit and rank, and appends it to self.cards.

Printing the Deck

Here is a __str__ method for Deck:

```
#inside class Deck:

    def __str__(self):
        res = []
        for card in self.cards:
            res.append(str(card))
        return '\n'.join(res)
```

This method demonstrates an efficient way to accumulate a large string: building a list of strings and then using join. The built-in function str invokes the __str__ method on each card and returns the string representation.

Since we invoke join on a newline character, the cards are separated by newlines. Here's what the result looks like:

```
>>> deck = Deck()
>>> print deck
Ace of Clubs
2 of Clubs
3 of Clubs
...
10 of Spades
Jack of Spades
Queen of Spades
King of Spades
```

Even though the result appears on 52 lines, it is one long string that contains newlines.

Add, Remove, Shuffle, and Sort

To deal cards, we would like a method that removes a card from the deck and returns it. The list method pop provides a convenient way to do that:

```
#inside class Deck:

    def pop_card(self):
        return self.cards.pop()
```

Since pop removes the *last* card in the list, we are dealing from the bottom of the deck. In real life "bottom dealing" is frowned upon, but in this context it's ok.

To add a card, we can use the list method append:

```
#inside class Deck:

    def add_card(self, card):
        self.cards.append(card)
```

A method like this that uses another function without doing much real work is sometimes called a **veneer**. The metaphor comes from woodworking, where it is common to glue a thin layer of good quality wood to the surface of a cheaper piece of wood.

In this case we are defining a "thin" method that expresses a list operation in terms that are appropriate for decks.

As another example, we can write a Deck method named shuffle using the function shuffle from the random module:

```
# inside class Deck:

    def shuffle(self):
        random.shuffle(self.cards)
```

Don't forget to import random.

Exercise 18-2.
Write a Deck method named sort that uses the list method sort to sort the cards in a Deck. sort uses the __cmp__ method we defined to determine sort order.

Inheritance

The language feature most often associated with object-oriented programming is **inheritance**. Inheritance is the ability to define a new class that is a modified version of an existing class.

It is called "inheritance" because the new class inherits the methods of the existing class. Extending this metaphor, the existing class is called the **parent** and the new class is called the **child**.

As an example, let's say we want a class to represent a "hand," that is, the set of cards held by one player. A hand is similar to a deck: both are made up of a set of cards, and both require operations like adding and removing cards.

A hand is also different from a deck; there are operations we want for hands that don't make sense for a deck. For example, in poker we might compare two hands to see which one wins. In bridge, we might compute a score for a hand in order to make a bid.

This relationship between classes—similar, but different—lends itself to inheritance.

The definition of a child class is like other class definitions, but the name of the parent class appears in parentheses:

```
class Hand(Deck):
    """Represents a hand of playing cards."""
```

This definition indicates that Hand inherits from Deck; that means we can use methods like pop_card and add_card for Hands as well as Decks.

Hand also inherits __init__ from Deck, but it doesn't really do what we want: instead of populating the hand with 52 new cards, the init method for Hands should initialize cards with an empty list.

If we provide an init method in the Hand class, it overrides the one in the Deck class:

```
# inside class Hand:

    def __init__(self, label=''):
        self.cards = []
        self.label = label
```

So when you create a Hand, Python invokes this init method:

```
>>> hand = Hand('new hand')
>>> print hand.cards
[]
>>> print hand.label
new hand
```

But the other methods are inherited from Deck, so we can use pop_card and add_card to deal a card:

```
>>> deck = Deck()
>>> card = deck.pop_card()
>>> hand.add_card(card)
>>> print hand
King of Spades
```

A natural next step is to encapsulate this code in a method called move_cards:

```
#inside class Deck:

    def move_cards(self, hand, num):
        for i in range(num):
            hand.add_card(self.pop_card())
```

move_cards takes two arguments, a Hand object and the number of cards to deal. It modifies both self and hand, and returns None.

In some games, cards are moved from one hand to another, or from a hand back to the deck. You can use move_cards for any of these operations: self can be either a Deck or a Hand, and hand, despite the name, can also be a Deck.

Exercise 18-3.
Write a Deck method called deal_hands that takes two parameters, the number of hands and the number of cards per hand, and that creates new Hand objects, deals the appropriate number of cards per hand, and returns a list of Hand objects.

Inheritance is a useful feature. Some programs that would be repetitive without inheritance can be written more elegantly with it. Inheritance can facilitate code reuse, since you can customize the behavior of parent classes without having to modify them. In some cases, the inheritance structure reflects the natural structure of the problem, which makes the program easier to understand.

On the other hand, inheritance can make programs difficult to read. When a method is invoked, it is sometimes not clear where to find its definition. The relevant code may be scattered among several modules. Also, many of the things that can be done using inheritance can be done as well or better without it.

Class Diagrams

So far we have seen stack diagrams, which show the state of a program, and object diagrams, which show the attributes of an object and their values. These diagrams represent a snapshot in the execution of a program, so they change as the program runs.

They are also highly detailed; for some purposes, too detailed. A class diagram is a more abstract representation of the structure of a program. Instead of showing individual objects, it shows classes and the relationships between them.

There are several kinds of relationship between classes:

- Objects in one class might contain references to objects in another class. For example, each Rectangle contains a reference to a Point, and each Deck contains references to many Cards. This kind of relationship is called **HAS-A**, as in, "a Rectangle has a Point."

- One class might inherit from another. This relationship is called **IS-A**, as in, "a Hand is a kind of a Deck."

- One class might depend on another in the sense that changes in one class would require changes in the other.

A **class diagram** is a graphical representation of these relationships. For example, Figure 18-2 shows the relationships between Card, Deck and Hand.

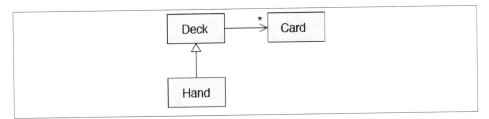

Figure 18-2. Class diagram.

The arrow with a hollow triangle head represents an IS-A relationship; in this case it indicates that Hand inherits from Deck.

The standard arrow head represents a HAS-A relationship; in this case a Deck has references to Card objects.

The star (*) near the arrow head is a **multiplicity**; it indicates how many Cards a Deck has. A multiplicity can be a simple number, like 52, a range, like 5..7 or a star, which indicates that a Deck can have any number of Cards.

A more detailed diagram might show that a Deck actually contains a *list* of Cards, but built-in types like list and dict are usually not included in class diagrams.

Exercise 18-4.

Read `TurtleWorld.py`, `World.py` and `Gui.py` and draw a class diagram that shows the relationships among the classes defined there.

Debugging

Inheritance can make debugging a challenge because when you invoke a method on an object, you might not know which method will be invoked.

Suppose you are writing a function that works with Hand objects. You would like it to work with all kinds of Hands, like PokerHands, BridgeHands, etc. If you invoke a method like `shuffle`, you might get the one defined in `Deck`, but if any of the subclasses override this method, you'll get that version instead.

Any time you are unsure about the flow of execution through your program, the simplest solution is to add print statements at the beginning of the relevant methods. If `Deck.shuffle` prints a message that says something like `Running Deck.shuffle`, then as the program runs it traces the flow of execution.

As an alternative, you could use this function, which takes an object and a method name (as a string) and returns the class that provides the definition of the method:

```
def find_defining_class(obj, meth_name):
    for ty in type(obj).mro():
        if meth_name in ty.__dict__:
            return ty
```

Here's an example:

```
>>> hand = Hand()
>>> print find_defining_class(hand, 'shuffle')
<class 'Card.Deck'>
```

So the `shuffle` method for this Hand is the one in `Deck`.

`find_defining_class` uses the `mro` method to get the list of class objects (types) that will be searched for methods. "MRO" stands for "method resolution order."

Here's a program design suggestion: whenever you override a method, the interface of the new method should be the same as the old. It should take the same parameters, return the same type, and obey the same preconditions and postconditions. If you obey this rule, you will find that any function designed to work with an instance of a super-class, like a Deck, will also work with instances of subclasses like a Hand or PokerHand.

If you violate this rule, your code will collapse like (sorry) a house of cards.

Data Encapsulation

Chapter 16 demonstrates a development plan we might call "object-oriented design." We identified objects we needed—Time, Point and Rectangle—and defined classes to represent them. In each case there is an obvious correspondence between the object and some entity in the real world (or at least a mathematical world).

But sometimes it is less obvious what objects you need and how they should interact. In that case you need a different development plan. In the same way that we discovered function interfaces by encapsulation and generalization, we can discover class interfaces by **data encapsulation**.

Markov analysis, from "Markov Analysis" (page 153), provides a good example. If you download my code from *http://thinkpython.com/code/markov.py*, you'll see that it uses two global variables—suffix_map and prefix—that are read and written from several functions.

```
suffix_map = {}
prefix = ()
```

Because these variables are global we can only run one analysis at a time. If we read two texts, their prefixes and suffixes would be added to the same data structures (which makes for some interesting generated text).

To run multiple analyses, and keep them separate, we can encapsulate the state of each analysis in an object. Here's what that looks like:

```
class Markov(object):

    def __init__(self):
        self.suffix_map = {}
        self.prefix = ()
```

Next, we transform the functions into methods. For example, here's process_word:

```
def process_word(self, word, order=2):
    if len(self.prefix) < order:
        self.prefix += (word,)
        return

    try:
        self.suffix_map[self.prefix].append(word)
    except KeyError:
        # if there is no entry for this prefix, make one
        self.suffix_map[self.prefix] = [word]

    self.prefix = shift(self.prefix, word)
```

Transforming a program like this—changing the design without changing the function—is another example of refactoring (see "Refactoring" (page 43)).

This example suggests a development plan for designing objects and methods:

1. Start by writing functions that read and write global variables (when necessary).
2. Once you get the program working, look for associations between global variables and the functions that use them.
3. Encapsulate related variables as attributes of an object.
4. Transform the associated functions into methods of the new class.

Exercise 18-5.
Download my code from "Markov Analysis" (page 153) (*http://thinkpython.com/code/ markov.py*), and follow the steps described above to encapsulate the global variables as attributes of a new class called Markov. Solution: *http://thinkpython.com/code/ Markov.py* (note the capital M).

Glossary

Encode:
> To represent one set of values using another set of values by constructing a mapping between them.

Class attribute:
> An attribute associated with a class object. Class attributes are defined inside a class definition but outside any method.

Instance attribute:
> An attribute associated with an instance of a class.

Veneer:
> A method or function that provides a different interface to another function without doing much computation.

Inheritance:
> The ability to define a new class that is a modified version of a previously defined class.

Parent class:
> The class from which a child class inherits.

Child class:
> A new class created by inheriting from an existing class; also called a "subclass."

IS-A relationship:
> The relationship between a child class and its parent class.

HAS-A relationship:
  The relationship between two classes where instances of one class contain references to instances of the other.

Class diagram:
  A diagram that shows the classes in a program and the relationships between them.

Multiplicity:
  A notation in a class diagram that shows, for a HAS-A relationship, how many references there are to instances of another class.

Exercises

Exercise 18-6.
The following are the possible hands in poker, in increasing order of value (and decreasing order of probability):

pair:
  two cards with the same rank

two pair:
  two pairs of cards with the same rank

three of a kind:
  three cards with the same rank

straight:
  five cards with ranks in sequence (aces can be high or low, so `Ace-2-3-4-5` is a straight and so is `10-Jack-Queen-King-Ace`, but `Queen-King-Ace-2-3` is not.)

flush:
  five cards with the same suit

full house:
  three cards with one rank, two cards with another

four of a kind:
  four cards with the same rank

straight flush:
  five cards in sequence (as defined above) and with the same suit

The goal of these exercises is to estimate the probability of drawing these various hands.

1. Download the following files from *http://thinkpython.com/code*:

   `Card.py`
     : A complete version of the `Card`, `Deck` and `Hand` classes in this chapter.

PokerHand.py
: An incomplete implementation of a class that represents a poker hand, and some code that tests it.

2. If you run PokerHand.py, it deals seven 7-card poker hands and checks to see if any of them contains a flush. Read this code carefully before you go on.

3. Add methods to PokerHand.py named has_pair, has_twopair, etc. that return True or False according to whether or not the hand meets the relevant criteria. Your code should work correctly for "hands" that contain any number of cards (although 5 and 7 are the most common sizes).

4. Write a method named classify that figures out the highest-value classification for a hand and sets the label attribute accordingly. For example, a 7-card hand might contain a flush and a pair; it should be labeled "flush."

5. When you are convinced that your classification methods are working, the next step is to estimate the probabilities of the various hands. Write a function in Poker Hand.py that shuffles a deck of cards, divides it into hands, classifies the hands, and counts the number of times various classifications appear.

6. Print a table of the classifications and their probabilities. Run your program with larger and larger numbers of hands until the output values converge to a reasonable degree of accuracy. Compare your results to the values at *http://en.wikipedia.org/wiki/Hand_rankings*.

Solution: *http://thinkpython.com/code/PokerHandSoln.py.*

Exercise 18-7.
This exercise uses TurtleWorld from Chapter 4. You will write code that makes Turtles play tag. If you are not familiar with the rules of tag, see *http://en.wikipedia.org/wiki/Tag_(game)*.

1. Download *http://thinkpython.com/code/Wobbler.py* and run it. You should see a TurtleWorld with three Turtles. If you press the Run button, the Turtles wander at random.

2. Read the code and make sure you understand how it works. The Wobbler class inherits from Turtle, which means that the Turtle methods lt, rt, fd and bk work on Wobblers.

   The step method gets invoked by TurtleWorld. It invokes steer, which turns the Turtle in the desired direction, wobble, which makes a random turn in proportion to the Turtle's clumsiness, and move, which moves forward a few pixels, depending on the Turtle's speed.

3. Create a file named Tagger.py. Import everything from Wobbler, then define a class named Tagger that inherits from Wobbler. Call make_world passing the Tagger class object as an argument.

4. Add a `steer` method to `Tagger` to override the one in `Wobbler`. As a starting place, write a version that always points the Turtle toward the origin. Hint: use the math function `atan2` and the Turtle attributes `x`, `y` and `heading`.

5. Modify `steer` so that the Turtles stay in bounds. For debugging, you might want to use the Step button, which invokes `step` once on each Turtle.

6. Modify `steer` so that each Turtle points toward its nearest neighbor. Hint: Turtles have an attribute, `world`, that is a reference to the TurtleWorld they live in, and the TurtleWorld has an attribute, `animals`, that is a list of all Turtles in the world.

7. Modify `steer` so the Turtles play tag. You can add methods to `Tagger` and you can override `steer` and `__init__`, but you may not modify or override `step`, `wobble` or `move`. Also, `steer` is allowed to change the heading of the Turtle but not the position.

   Adjust the rules and your `steer` method for good quality play; for example, it should be possible for the slow Turtle to tag the faster Turtles eventually.

Solution: *http://thinkpython.com/code/Tagger.py*.

## CHAPTER 19

Case Study: Tkinter

GUI

Most of the programs we have seen so far are text-based, but many programs use **graphical user interfaces**, also known as **GUIs**.

Python provides several choices for writing GUI-based programs, including wxPython, Tkinter, and Qt. Each has pros and cons, which is why Python has not converged on a standard.

The one I will present in this chapter is Tkinter because I think it is the easiest to get started with. Most of the concepts in this chapter apply to the other GUI modules, too.

There are several books and web pages about Tkinter. One of the best online resources is *An Introduction to Tkinter* by Fredrik Lundh.

I have written a module called Gui.py that comes with Swampy. It provides a simplified interface to the functions and classes in Tkinter. The examples in this chapter are based on this module.

Here is a simple example that creates and displays a Gui:

To create a GUI, you have to import Gui and instantiate a Gui object:

```
from Gui import *

g = Gui()
g.title('Gui')
g.mainloop()
```

When you run this code, a window should appear with an empty gray square and the title Gui. mainloop runs the **event loop**, which waits for the user to do something and responds accordingly. It is an infinite loop; it runs until the user closes the window, or presses Control-C, or does something that causes the program to quit.

This Gui doesn't do much because it doesn't have any **widgets**. Widgets are the elements that make up a GUI; they include:

Button:
> A widget, containing text or an image, that performs an action when pressed.

Canvas:
> A region that can display lines, rectangles, circles and other shapes.

Entry:
> A region where users can type text.

Scrollbar:
> A widget that controls the visible part of another widget.

Frame:
> A container, often invisible, that contains other widgets.

The empty gray square you see when you create a Gui is a Frame. When you create a new widget, it is added to this Frame.

Buttons and Callbacks

The method bu creates a Button widget:

```
button = g.bu(text='Press me.')
```

The return value from bu is a Button object. The button that appears in the Frame is a graphical representation of this object; you can control the button by invoking methods on it.

bu takes up to 32 parameters that control the appearance and function of the button. These parameters are called **options**. Instead of providing values for all 32 options, you can use keyword arguments, like text='Press me.', to specify only the options you need and use the default values for the rest.

When you add a widget to the Frame, it gets "shrink-wrapped;" that is, the Frame shrinks to the size of the Button. If you add more widgets, the Frame grows to accommodate them.

The method la creates a Label widget:

```
label = g.la(text='Press the button.')
```

By default, Tkinter stacks the widgets top-to-bottom and centers them. We'll see how to override that behavior soon.

If you press the button, you will see that it doesn't do much. That's because you haven't "wired it up;" that is, you haven't told it what to do!

The option that controls the behavior of a button is command. The value of command is a function that gets executed when the button is pressed. For example, here is a function that creates a new Label:

```
def make_label():
    g.la(text='Thank you.')
```

Now we can create a button with this function as its command:

```
button2 = g.bu(text='No, press me!', command=make_label)
```

When you press this button, it should execute make_label and a new label should appear.

The value of the command option is a function object, which is known as a **callback** because after you call bu to create the button, the flow of execution "calls back" when the user presses the button.

This kind of flow is characteristic of **event-driven programming**. User actions, like button presses and key strokes, are called **events**. In event-driven programming, the flow of execution is determined by user actions rather than by the programmer.

The challenge of event-driven programming is to construct a set of widgets and callbacks that work correctly (or at least generate appropriate error messages) for any sequence of user actions.

Exercise 19-1.
Write a program that creates a GUI with a single button. When the button is pressed it should create a second button. When *that* button is pressed, it should create a label that says, "Nice job!"

What happens if you press the buttons more than once? Solution: *http://thinkpython.com/code/button_demo.py*

Canvas Widgets

One of the most versatile widgets is the Canvas, which creates a region for drawing lines, circles and other shapes. If you did Exercise 15-4 you are already familiar with canvases.

The method ca creates a new Canvas:

```
canvas = g.ca(width=500, height=500)
```

width and height are the dimensions of the canvas in pixels.

After you create a widget, you can still change the values of the options with the con fig method. For example, the bg option changes the background color:

```
canvas.config(bg='white')
```

The value of bg is a string that names a color. The set of legal color names is different for different implementations of Python, but all implementations provide at least:

```
white    black
 red     green    blue
 cyan    yellow   magenta
```

Shapes on a Canvas are called **items**. For example, the Canvas method `circle` draws (you guessed it) a circle:

```
item = canvas.circle([0,0], 100, fill='red')
```

The first argument is a coordinate pair that specifies the center of the circle; the second is the radius.

`Gui.py` provides a standard Cartesian coordinate system with the origin at the center of the Canvas and the positive *y* axis pointing up. This is different from some other graphics systems where the origin is in the upper left corner, with the *y* axis pointing down.

The `fill` option specifies that the circle should be filled in with red.

The return value from `circle` is an Item object that provides methods for modifying the item on the canvas. For example, you can use `config` to change any of the circle's options:

```
item.config(fill='yellow', outline='orange', width=10)
```

`width` is the thickness of the outline in pixels; `outline` is the color.

Exercise 19-2.

Write a program that creates a Canvas and a Button. When the user presses the Button, it should draw a circle on the canvas.

Coordinate Sequences

The `rectangle` method takes a sequence of coordinates that specify opposite corners of the rectangle. This example draws a green rectangle with the lower left corner at the origin and the upper right corner at *(200,100)*:

```
canvas.rectangle([[0, 0], [200, 100]],
                  fill='blue', outline='orange', width=10)
```

This way of specifying corners is called a **bounding box** because the two points bound the rectangle.

`oval` takes a bounding box and draws an oval within the specified rectangle:

```
canvas.oval([[0, 0], [200, 100]], outline='orange', width=10)
```

`line` takes a sequence of coordinates and draws a line that connects the points. This example draws two legs of a triangle:

```
canvas.line([[0, 100], [100, 200], [200, 100]], width=10)
```

---

polygon takes the same arguments, but it draws the last leg of the polygon (if necessary) and fills it in:

```
canvas.polygon([[0, 100], [100, 200], [200, 100]],
               fill='red', outline='orange', width=10)
```

More Widgets

Tkinter provides two widgets that let users type text: an Entry, which is a single line, and a Text widget, which has multiple lines.

en creates a new Entry:

```
entry = g.en(text='Default text.')
```

The text option allows you to put text into the entry when it is created. The get method returns the contents of the Entry (which may have been changed by the user):

```
>>> entry.get()
'Default text.'
```

te creates a Text widget:

```
text = g.te(width=100, height=5)
```

width and height are the dimensions of the widget in characters and lines.

insert puts text into the Text widget:

```
text.insert(END, 'A line of text.')
```

END is a special index that indicates the last character in the Text widget.

You can also specify a character using a dotted index, like 1.1, which has the line number before the dot and the column number after. The following example adds the letters 'nother' after the first character of the first line.

```
>>> text.insert(1.1, 'nother')
```

The get method reads the text in the widget; it takes a start and end index as arguments. The following example returns all the text in the widget, including the newline character:

```
>>> text.get(0.0, END)
'Another line of text.\n'
```

The delete method removes text from the widget; the following example deletes all but the first two characters:

```
>>> text.delete(1.2, END)
>>> text.get(0.0, END)
'An\n'
```

Exercise 19-3.

Modify your solution to Exercise 19-2 by adding an Entry widget and a second button. When the user presses the second button, it should read a color name from the Entry and use it to change the fill color of the circle. Use `config` to modify the existing circle; don't create a new one.

Your program should handle the case where the user tries to change the color of a circle that hasn't been created, and the case where the color name is invalid.

You can see my solution at *http://thinkpython.com/code/circle_demo.py*.

Packing Widgets

So far we have been stacking widgets in a single column, but in most GUIs the layout is more complicated. For example, Figure 19-1 shows a simplified version of TurtleWorld (see Chapter 4).

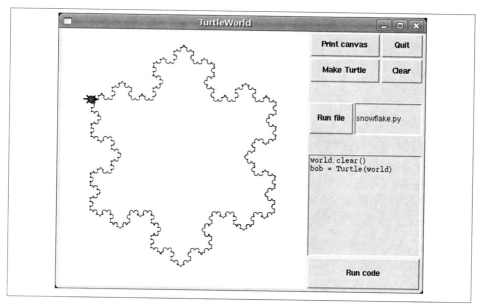

Figure 19-1. Class diagram.

This section presents the code that creates this GUI, broken into a series of steps. You can download the complete example from here (*http://thinkpython.com/code/Simple-TurtleWorld.py*).

At the top level, this GUI contains two widgets—a Canvas and a Frame—arranged in a row. So the first step is to create the row.

```
class SimpleTurtleWorld(TurtleWorld):
    """This class is identical to TurtleWorld, but the code that
    lays out the GUI is simplified for explanatory purposes."""

    def setup(self):
        self.row()
        ...
```

setup is the function that creates and arranges the widgets. Arranging widgets in a GUI is called **packing**.

row creates a row Frame and makes it the "current Frame." Until this Frame is closed or another Frame is created, all subsequent widgets are packed in a row.

Here is the code that creates the Canvas and the column Frame that hold the other widgets:

```
self.canvas = self.ca(width=400, height=400, bg='white')
self.col()
```

The first widget in the column is a grid Frame, which contains four buttons arranged two-by-two:

```
self.gr(cols=2)
self.bu(text='Print canvas', command=self.canvas.dump)
self.bu(text='Quit', command=self.quit)
self.bu(text='Make Turtle', command=self.make_turtle)
self.bu(text='Clear', command=self.clear)
self.endgr()
```

gr creates the grid; the argument is the number of columns. Widgets in the grid are laid out left-to-right, top-to-bottom.

The first button uses self.canvas.dump as a callback; the second uses self.quit. These are **bound methods**, which means they are associated with a particular object. When they are invoked, they are invoked on the object.

The next widget in the column is a row Frame that contains a Button and an Entry:

```
self.row([0,1], pady=30)
self.bu(text='Run file', command=self.run_file)
self.en_file = self.en(text='snowflake.py', width=5)
self.endrow()
```

The first argument to row is a list of weights that determines how extra space is allocated between widgets. The list [0,1] means that all extra space is allocated to the second widget, which is the Entry. If you run this code and resize the window, you will see that the Entry grows and the Button doesn't.

The option pady "pads" this row in the y direction, adding 30 pixels of space above and below.

endrow ends this row of widgets, so subsequent widgets are packed in the column Frame. Gui.py keeps a stack of Frames:

- When you use row, col or gr to create a Frame, it goes on top of the stack and becomes the current Frame.

- When you use endrow, endcol or endgr to close a Frame, it gets popped off the stack and the previous Frame on the stack becomes the current Frame.

The method run_file reads the contents of the Entry, uses it as a filename, reads the contents and passes it to run_code. self.inter as an Interpreter object that knows how to take a string and execute it as Python code.

```
def run_file(self):
    filename = self.en_file.get()
    fp = open(filename)
    source = fp.read()
    self.inter.run_code(source, filename)
```

The last two widgets are a Text widget and a Button:

```
self.te_code = self.te(width=25, height=10)
self.te_code.insert(END, 'world.clear()\n')
self.te_code.insert(END, 'bob = Turtle(world)\n')

self.bu(text='Run code', command=self.run_text)
```

run_text is similar to run_file except that it takes the code from the Text widget instead of from a file:

```
def run_text(self):
    source = self.te_code.get(1.0, END)
    self.inter.run_code(source, '<user-provided code>')
```

Unfortunately, the details of widget layout are different in other languages, and in different Python modules. Tkinter alone provides three different mechanisms for arranging widgets. These mechanisms are called **geometry managers**. The one I demonstrated in this section is the "grid" geometry manager; the others are called "pack" and "place."

Fortunately, most of the concepts in this section apply to other GUI modules and other languages.

Menus and Callables

A Menubutton is a widget that looks like a button, but when pressed it pops up a menu. After the user selects an item, the menu disappears.

Here is code that creates a color selection Menubutton (you can download it from *http://thinkpython.com/code/menubutton_demo.py*):

```
g = Gui()
g.la('Select a color:')
colors = ['red', 'green', 'blue']
mb = g.mb(text=colors[0])
```

mb creates the Menubutton. Initially, the text on the button is the name of the default color. The following loop creates one menu item for each color:

```
for color in colors:
    g.mi(mb, text=color, command=Callable(set_color, color))
```

The first argument of mi is the Menubutton these items are associated with.

The command option is a Callable object, which is something new. So far we have seen functions and bound methods used as callbacks, which works fine if you don't have to pass any arguments to the function. Otherwise you have to construct a Callable object that contains a function, like set_color, and its arguments, like color.

The Callable object stores a reference to the function and the arguments as attributes. Later, when the user clicks on a menu item, the callback calls the function and passes the stored arguments.

Here is what set_color might look like:

```
def set_color(color):
    mb.config(text=color)
    print color
```

When the user selects a menu item and set_color is called, it configures the Menu-button to display the newly-selected color. It also print the color; if you try this example, you can confirm that set_color is called when you select an item (and *not* called when you create the Callable object).

Binding

A **binding** is an association between a widget, an event and a callback: when an event (like a button press) happens on a widget, the callback is invoked.

Many widgets have default bindings. For example, when you press a button, the default binding changes the relief of the button to make it look depressed. When you release the button, the binding restores the appearance of the button and invokes the callback specified with the command option.

You can use the bind method to override these default bindings or to add new ones. For example, this code creates a binding for a canvas (you can download the code in this section from *http://thinkpython.com/code/draggable_demo.py*):

```
ca.bind('<ButtonPress-1>', make_circle)
```

The first argument is an event string; this event is triggered when the user presses the left mouse button. Other mouse events include ButtonMotion, ButtonRelease and Double-Button.

The second argument is an event handler. An event handler is a function or bound method, like a callback, but an important difference is that an event handler takes an Event object as a parameter. Here is an example:

```
def make_circle(event):
    pos = ca.canvas_coords([event.x, event.y])
    item = ca.circle(pos, 5, fill='red')
```

The Event object contains information about the type of event and details like the coordinates of the mouse pointer. In this example the information we need is the location of the mouse click. These values are in "pixel coordinates," which are defined by the underlying graphical system. The method canvas_coords translates them to "Canvas coordinates," which are compatible with Canvas methods like circle.

For Entry widgets, it is common to bind the <Return> event, which is triggered when the user presses the Return or Enter key. For example, the following code creates a Button and an Entry.

```
bu = g.bu('Make text item:', make_text)
en = g.en()
en.bind('<Return>', make_text)
```

make_text is called when the Button is pressed or when the user hits Return while typing in the Entry. To make this work, we need a function that can be called as a command (with no arguments) or as an event handler (with an Event as an argument):

```
def make_text(event=None):
    text = en.get()
    item = ca.text([0,0], text)
```

make_text gets the contents of the Entry and displays it as a Text item in the Canvas.

It is also possible to create bindings for Canvas items. The following is a class definition for Draggable, which is a child class of Item that provides bindings that implement drag-and-drop capability.

```
class Draggable(Item):

    def __init__(self, item):
        self.canvas = item.canvas
        self.tag = item.tag
        self.bind('<Button-3>', self.select)
        self.bind('<B3-Motion>', self.drag)
        self.bind('<Release-3>', self.drop)
```

The init method takes an Item as a parameter. It copies the attributes of the Item and then creates bindings for three events: a button press, button motion, and button release.

The event handler `select` stores the coordinates of the current event and the original color of the item, then changes the color to yellow:

```
def select(self, event):
    self.dragx = event.x
    self.dragy = event.y

    self.fill = self.cget('fill')
    self.config(fill='yellow')
```

`cget` stands for "get configuration;" it takes the name of an option as a string and returns the current value of that option.

`drag` computes how far the object has moved relative to the starting place, updates the stored coordinates, and then moves the item.

```
def drag(self, event):
    dx = event.x - self.dragx
    dy = event.y - self.dragy

    self.dragx = event.x
    self.dragy = event.y

    self.move(dx, dy)
```

This computation is done in pixel coordinates; there is no need to convert to Canvas coordinates.

Finally, `drop` restores the original color of the item:

```
def drop(self, event):
    self.config(fill=self.fill)
```

You can use the `Draggable` class to add drag-and-drop capability to an existing item. For example, here is a modified version of `make_circle` that uses `circle` to create an Item and `Draggable` to make it draggable:

```
def make_circle(event):
    pos = ca.canvas_coords([event.x, event.y])
    item = ca.circle(pos, 5, fill='red')
    item = Draggable(item)
```

This example demonstrates one of the benefits of inheritance: you can modify the capabilities of a parent class without modifying its definition. This is particularly useful if you want to change behavior defined in a module you did not write.

Debugging

One of the challenges of GUI programming is keeping track of which things happen while the GUI is being built and which things happen later in response to user events.

For example, when you are setting up a callback, it is a common error to call the function rather than passing a reference to it:

```
def the_callback():
    print 'Called.'

g.bu(text='This is wrong!', command=the_callback())
```

If you run this code, you will see that it calls `the_callback` immediately, and *then* creates the button. When you press the button, it does nothing because the return value from `the_callback` is None. Usually you do not want to invoke a callback while you are setting up the GUI; it should only be invoked later in response to a user event.

Another challenge of GUI programming is that you don't have control of the flow of execution. Which parts of the program execute and their order are determined by user actions. That means that you have to design your program to work correctly for any possible sequence of events.

For example, the GUI in Exercise 19-3 has two widgets: one creates a Circle item and the other changes the color of the Circle. If the user creates the circle and then changes the color, there's no problem. But what if the user changes the color of a circle that doesn't exist yet? Or creates more than one circle?

As the number of widgets grows, it is increasingly difficult to imagine all possible sequences of events. One way to manage this complexity is to encapsulate the state of the system in an object and then consider:

- What are the possible states? In the Circle example, we might consider two states: before and after the user creates the first circle.

- In each state, what events can occur? In the example, the user can press either of the buttons, or quit.

- For each state-event pair, what is the desired outcome? Since there are two states and two buttons, there are four state-event pairs to consider.

- What can cause a transition from one state to another? In this case, there is a transition when the user creates the first circle.

You might also find it useful to define, and check, invariants that should hold regardless of the sequence of events.

This approach to GUI programming can help you write correct code without taking the time to test every possible sequence of user events!

Glossary

GUI:
    A graphical user interface.

Widget:
    One of the elements that makes up a GUI: buttons, menus, text entry fields, etc.

Option:
    A value that controls the appearance or function of a widget.

Keyword argument:
    An argument that indicates the parameter name as part of the function call.

Callback:
    A function associated with a widget that is called when the user performs an action.

Bound method:
    A method associated with a particular instance.

Event-driven programming:
    A style of programming in which the flow of execution is determined by user actions.

Event:
    A user action, like a mouse click or key press, that causes a GUI to respond.

Event loop:
    An infinite loop that waits for user actions and responds.

Item:
    A graphical element on a Canvas widget.

Bounding box:
    A rectangle that encloses a set of items, usually specified by two opposing corners.

Pack:
    To arrange and display the elements of a GUI.

Geometry manager:
    A system for packing widgets.

Binding:
    An association between a widget, an event, and an event handler. The event handler is called when the event occurs in the widget.

Exercises

Exercise 19-4.

For this exercise, you will write an image viewer. Here is a simple example:

```
g = Gui()
canvas = g.ca(width=300)
photo = PhotoImage(file='danger.gif')
canvas.image([0,0], image=photo)
g.mainloop()
```

`PhotoImage` reads a file and returns a `PhotoImage` object that Tkinter can display. `Canvas.image` puts the image on the canvas, centered on the given coordinates. You can also put images on labels, buttons, and some other widgets:

```
g.la(image=photo)
g.bu(image=photo)
```

`PhotoImage` can only handle a few image formats, like GIF and PPM, but we can use the Python Imaging Library (PIL) to read other files.

The name of the PIL module is `Image`, but Tkinter defines an object with the same name. To avoid the conflict, you can use `import...as` like this:

```
import Image as PIL
import ImageTk
```

The first line imports `Image` and gives it the local name `PIL`. The second line imports `ImageTk`, which can translate a PIL image into a Tkinter PhotoImage. Here's an example:

```
image = PIL.open('allen.png')
photo2 = ImageTk.PhotoImage(image)
g.la(image=photo2)
```

1. Download `image_demo.py`, `danger.gif` and `allen.png` from *http://thinkpython.com/code*. Run `image_demo.py`. You might have to install `PIL` and `ImageTk`. They are probably in your software repository, but if not you can get them from *http://pythonware.com/products/pil/*.

2. In `image_demo.py` change the name of the second PhotoImage from `photo2` to `photo` and run the program again. You should see the second PhotoImage but not the first.

   The problem is that when you reassign `photo` it overwrites the reference to the first PhotoImage, which then disappears. The same thing happens if you assign a PhotoImage to a local variable; it disappears when the function ends.

   To avoid this problem, you have to store a reference to each PhotoImage you want to keep. You can use a global variable, or store PhotoImages in a data structure or as an attribute of an object.

This behavior can be frustrating, which is why I am warning you (and why the example image says "Danger!").

3. Starting with this example, write a program that takes the name of a directory and loops through all the files, displaying any files that PIL recognizes as images. You can use a `try` statement to catch the files PIL doesn't recognize.

   When the user clicks on the image, the program should display the next one.

4. PIL provides a variety of methods for manipulating images. You can read about them at *http://pythonware.com/library/pil/handbook*. As a challenge, choose a few of these methods and provide a GUI for applying them to images.

Solution: *http://thinkpython.com/code/ImageBrowser.py*.

Exercise 19-5.
A vector graphics editor is a program that allows users to draw and edit shapes on the screen and generate output files in vector graphics formats like Postscript and SVG.

Write a simple vector graphics editor using Tkinter. At a minimum, it should allow users to draw lines, circles and rectangles, and it should use `Canvas.dump` to generate a Postscript description of the contents of the Canvas.

As a challenge, you could allow users to select and resize items on the Canvas.

Exercise 19-6.
Use Tkinter to write a basic web browser. It should have a Text widget where the user can enter a URL and a Canvas to display the contents of the page.

You can use the `urllib` module to download files (see Exercise 14-6) and you can use the `HTMLParser` module to parse the HTML tags (see *http://docs.python.org/lib/module-HTMLParser.html*).

At a minimum your browser should handle plain text and hyperlinks. As a challenge you could handle background colors, text formatting tags and images.

Debugging

Different kinds of errors can occur in a program, and it is useful to distinguish among them in order to track them down more quickly:

- Syntax errors are produced by Python when it is translating the source code into byte code. They usually indicate that there is something wrong with the syntax of the program. Example: Omitting the colon at the end of a `def` statement yields the somewhat redundant message `SyntaxError: invalid syntax`.

- Runtime errors are produced by the interpreter if something goes wrong while the program is running. Most runtime error messages include information about where the error occurred and what functions were executing. Example: An infinite recursion eventually causes the runtime error "maximum recursion depth exceeded."

- Semantic errors are problems with a program that runs without producing error messages but doesn't do the right thing. Example: An expression may not be evaluated in the order you expect, yielding an incorrect result.

The first step in debugging is to figure out which kind of error you are dealing with. Although the following sections are organized by error type, some techniques are applicable in more than one situation.

Syntax Errors

Syntax errors are usually easy to fix once you figure out what they are. Unfortunately, the error messages are often not helpful. The most common messages are `SyntaxError: invalid syntax` and `SyntaxError: invalid token`, neither of which is very informative.

On the other hand, the message does tell you where in the program the problem occurred. Actually, it tells you where Python noticed a problem, which is not necessarily where the error is. Sometimes the error is prior to the location of the error message, often on the preceding line.

If you are building the program incrementally, you should have a good idea about where the error is. It will be in the last line you added.

If you are copying code from a book, start by comparing your code to the book's code very carefully. Check every character. At the same time, remember that the book might be wrong, so if you see something that looks like a syntax error, it might be.

Here are some ways to avoid the most common syntax errors:

1. Make sure you are not using a Python keyword for a variable name.

2. Check that you have a colon at the end of the header of every compound statement, including `for`, `while`, `if`, and `def` statements.

3. Make sure that any strings in the code have matching quotation marks.

4. If you have multiline strings with triple quotes (single or double), make sure you have terminated the string properly. An unterminated string may cause an `invalid token` error at the end of your program, or it may treat the following part of the program as a string until it comes to the next string. In the second case, it might not produce an error message at all!

5. An unclosed opening operator—(, {, or [—makes Python continue with the next line as part of the current statement. Generally, an error occurs almost immediately in the next line.

6. Check for the classic = instead of == inside a conditional.

7. Check the indentation to make sure it lines up the way it is supposed to. Python can handle space and tabs, but if you mix them it can cause problems. The best way to avoid this problem is to use a text editor that knows about Python and generates consistent indentation.

If nothing works, move on to the next section...

I Keep Making Changes and It Makes No Difference

If the interpreter says there is an error and you don't see it, that might be because you and the interpreter are not looking at the same code. Check your programming environment to make sure that the program you are editing is the one Python is trying to run.

If you are not sure, try putting an obvious and deliberate syntax error at the beginning of the program. Now run it again. If the interpreter doesn't find the new error, you are not running the new code.

There are a few likely culprits:

- You edited the file and forgot to save the changes before running it again. Some programming environments do this for you, but some don't.
- You changed the name of the file, but you are still running the old name.
- Something in your development environment is configured incorrectly.
- If you are writing a module and using `import`, make sure you don't give your module the same name as one of the standard Python modules.
- If you are using `import` to read a module, remember that you have to restart the interpreter or use `reload` to read a modified file. If you import the module again, it won't do anything.

If you get stuck and you can't figure out what is going on, one approach is to start again with a new program like "Hello, World!," and make sure you can get a known program to run. Then gradually add the pieces of the original program to the new one.

Runtime Errors

Once your program is syntactically correct, Python can compile it and at least start running it. What could possibly go wrong?

My Program Does Absolutely Nothing

This problem is most common when your file consists of functions and classes but does not actually invoke anything to start execution. This may be intentional if you only plan to import this module to supply classes and functions.

If it is not intentional, make sure that you are invoking a function to start execution, or execute one from the interactive prompt. Also see the "Flow of Execution" section below.

My Program Hangs

If a program stops and seems to be doing nothing, it is "hanging." Often that means that it is caught in an infinite loop or infinite recursion.

- If there is a particular loop that you suspect is the problem, add a `print` statement immediately before the loop that says "entering the loop" and another immediately after that says "exiting the loop."

  Run the program. If you get the first message and not the second, you've got an infinite loop. Go to the "Infinite Loop" section below.

- Most of the time, an infinite recursion will cause the program to run for a while and then produce a "RuntimeError: Maximum recursion depth exceeded" error. If that happens, go to the "Infinite Recursion" section below.

If you are not getting this error but you suspect there is a problem with a recursive method or function, you can still use the techniques in the "Infinite Recursion" section.

- If neither of those steps works, start testing other loops and other recursive functions and methods.

- If that doesn't work, then it is possible that you don't understand the flow of execution in your program. Go to the "Flow of Execution" section below.

Infinite loop

If you think you have an infinite loop and you think you know what loop is causing the problem, add a print statement at the end of the loop that prints the values of the variables in the condition and the value of the condition.

For example:

```
while x > 0 and y < 0 :
    # do something to x
    # do something to y

    print "x: ", x
    print "y: ", y
    print "condition: ", (x > 0 and y < 0)
```

Now when you run the program, you will see three lines of output for each time through the loop. The last time through the loop, the condition should be false. If the loop keeps going, you will be able to see the values of x and y, and you might figure out why they are not being updated correctly.

Infinite recursion

Most of the time, an infinite recursion will cause the program to run for a while and then produce a Maximum recursion depth exceeded error.

If you suspect that a function or method is causing an infinite recursion, start by checking to make sure that there is a base case. In other words, there should be some condition that will cause the function or method to return without making a recursive invocation. If not, then you need to rethink the algorithm and identify a base case.

If there is a base case but the program doesn't seem to be reaching it, add a print statement at the beginning of the function or method that prints the parameters. Now when you run the program, you will see a few lines of output every time the function or method is invoked, and you will see the parameters. If the parameters are not moving toward the base case, you will get some ideas about why not.

Flow of execution

If you are not sure how the flow of execution is moving through your program, add `print` statements to the beginning of each function with a message like "entering function `foo`," where `foo` is the name of the function.

Now when you run the program, it will print a trace of each function as it is invoked.

When I Run the Program, I Get an Exception

If something goes wrong during runtime, Python prints a message that includes the name of the exception, the line of the program where the problem occurred, and a traceback.

The traceback identifies the function that is currently running, and then the function that invoked it, and then the function that invoked *that*, and so on. In other words, it traces the sequence of function invocations that got you to where you are. It also includes the line number in your file where each of these calls occurs.

The first step is to examine the place in the program where the error occurred and see if you can figure out what happened. These are some of the most common runtime errors:

NameError:
> You are trying to use a variable that doesn't exist in the current environment. Remember that local variables are local. You cannot refer to them from outside the function where they are defined.

TypeError:
> There are several possible causes:

> - You are trying to use a value improperly. Example: indexing a string, list, or tuple with something other than an integer.

> - There is a mismatch between the items in a format string and the items passed for conversion. This can happen if either the number of items does not match or an invalid conversion is called for.

> - You are passing the wrong number of arguments to a function or method. For methods, look at the method definition and check that the first parameter is `self`. Then look at the method invocation; make sure you are invoking the method on an object with the right type and providing the other arguments correctly.

KeyError:
> You are trying to access an element of a dictionary using a key that the dictionary does not contain.

AttributeError:

You are trying to access an attribute or method that does not exist. Check the spelling! You can use `dir` to list the attributes that do exist.

If an AttributeError indicates that an object has `NoneType`, that means that it is `None`. One common cause is forgetting to return a value from a function; if you get to the end of a function without hitting a `return` statement, it returns `None`. Another common cause is using the result from a list method, like `sort`, that returns `None`.

IndexError:

The index you are using to access a list, string, or tuple is greater than its length minus one. Immediately before the site of the error, add a `print` statement to display the value of the index and the length of the array. Is the array the right size? Is the index the right value?

The Python debugger (`pdb`) is useful for tracking down Exceptions because it allows you to examine the state of the program immediately before the error. You can read about `pdb` at *http://docs.python.org/lib/module-pdb.html*.

I Added So Many Print Statements I Get Inundated with Output

One of the problems with using `print` statements for debugging is that you can end up buried in output. There are two ways to proceed: simplify the output or simplify the program.

To simplify the output, you can remove or comment out `print` statements that aren't helping, or combine them, or format the output so it is easier to understand.

To simplify the program, there are several things you can do. First, scale down the problem the program is working on. For example, if you are searching a list, search a *small* list. If the program takes input from the user, give it the simplest input that causes the problem.

Second, clean up the program. Remove dead code and reorganize the program to make it as easy to read as possible. For example, if you suspect that the problem is in a deeply nested part of the program, try rewriting that part with simpler structure. If you suspect a large function, try splitting it into smaller functions and testing them separately.

Often the process of finding the minimal test case leads you to the bug. If you find that a program works in one situation but not in another, that gives you a clue about what is going on.

Similarly, rewriting a piece of code can help you find subtle bugs. If you make a change that you think shouldn't affect the program, and it does, that can tip you off.

Semantic Errors

In some ways, semantic errors are the hardest to debug, because the interpreter provides no information about what is wrong. Only you know what the program is supposed to do.

The first step is to make a connection between the program text and the behavior you are seeing. You need a hypothesis about what the program is actually doing. One of the things that makes that hard is that computers run so fast.

You will often wish that you could slow the program down to human speed, and with some debuggers you can. But the time it takes to insert a few well-placed `print` statements is often short compared to setting up the debugger, inserting and removing breakpoints, and "stepping" the program to where the error is occurring.

My Program Doesn't Work

You should ask yourself these questions:

- Is there something the program was supposed to do but which doesn't seem to be happening? Find the section of the code that performs that function and make sure it is executing when you think it should.

- Is something happening that shouldn't? Find code in your program that performs that function and see if it is executing when it shouldn't.

- Is a section of code producing an effect that is not what you expected? Make sure that you understand the code in question, especially if it involves invocations to functions or methods in other Python modules. Read the documentation for the functions you invoke. Try them out by writing simple test cases and checking the results.

In order to program, you need to have a mental model of how programs work. If you write a program that doesn't do what you expect, very often the problem is not in the program; it's in your mental model.

The best way to correct your mental model is to break the program into its components (usually the functions and methods) and test each component independently. Once you find the discrepancy between your model and reality, you can solve the problem.

Of course, you should be building and testing components as you develop the program. If you encounter a problem, there should be only a small amount of new code that is not known to be correct.

I've Got a Big Hairy Expression and It Doesn't Do What I Expect

Writing complex expressions is fine as long as they are readable, but they can be hard to debug. It is often a good idea to break a complex expression into a series of assignments to temporary variables.

For example:

```
self.hands[i].addCard(self.hands[self.findNeighbor(i)].popCard())
```

This can be rewritten as:

```
neighbor = self.findNeighbor(i)
pickedCard = self.hands[neighbor].popCard()
self.hands[i].addCard(pickedCard)
```

The explicit version is easier to read because the variable names provide additional documentation, and it is easier to debug because you can check the types of the intermediate variables and display their values.

Another problem that can occur with big expressions is that the order of evaluation may not be what you expect. For example, if you are translating the expression $\frac{x}{2\pi}$ into Python, you might write:

```
y = x / 2 * math.pi
```

That is not correct because multiplication and division have the same precedence and are evaluated from left to right. So this expression computes $x\pi / 2$.

A good way to debug expressions is to add parentheses to make the order of evaluation explicit:

```
y = x / (2 * math.pi)
```

Whenever you are not sure of the order of evaluation, use parentheses. Not only will the program be correct (in the sense of doing what you intended), it will also be more readable for other people who haven't memorized the rules of precedence.

I've Got a Function or Method That Doesn't Return What I Expect

If you have a `return` statement with a complex expression, you don't have a chance to print the `return` value before returning. Again, you can use a temporary variable. For example, instead of:

```
return self.hands[i].removeMatches()
```

you could write:

```
count = self.hands[i].removeMatches()
return count
```

Now you have the opportunity to display the value of `count` before returning.

I'm Really, Really Stuck and I Need Help

First, try getting away from the computer for a few minutes. Computers emit waves that affect the brain, causing these symptoms:

- Frustration and rage.
- Superstitious beliefs ("the computer hates me") and magical thinking ("the program only works when I wear my hat backward").
- Random walk programming (the attempt to program by writing every possible program and choosing the one that does the right thing).

If you find yourself suffering from any of these symptoms, get up and go for a walk. When you are calm, think about the program. What is it doing? What are some possible causes of that behavior? When was the last time you had a working program, and what did you do next?

Sometimes it just takes time to find a bug. I often find bugs when I am away from the computer and let my mind wander. Some of the best places to find bugs are trains, showers, and in bed, just before you fall asleep.

No, I Really Need Help

It happens. Even the best programmers occasionally get stuck. Sometimes you work on a program so long that you can't see the error. A fresh pair of eyes is just the thing.

Before you bring someone else in, make sure you are prepared. Your program should be as simple as possible, and you should be working on the smallest input that causes the error. You should have `print` statements in the appropriate places (and the output they produce should be comprehensible). You should understand the problem well enough to describe it concisely.

When you bring someone in to help, be sure to give them the information they need:

- If there is an error message, what is it and what part of the program does it indicate?
- What was the last thing you did before this error occurred? What were the last lines of code that you wrote, or what is the new test case that fails?
- What have you tried so far, and what have you learned?

When you find the bug, take a second to think about what you could have done to find it faster. Next time you see something similar, you will be able to find the bug more quickly.

Remember, the goal is not just to make the program work. The goal is to learn how to make the program work.

Analysis of Algorithms

This appendix is an edited excerpt from *Think Complexity*, by Allen B. Downey, also published by O'Reilly Media (2011). When you are done with this book, you might want to move on to that one.

Analysis of algorithms is a branch of computer science that studies the performance of algorithms, especially their run time and space requirements. See *http://en.wikipedia.org/wiki/Analysis_of_algorithms*.

The practical goal of algorithm analysis is to predict the performance of different algorithms in order to guide design decisions.

During the 2008 United States Presidential Campaign, candidate Barack Obama was asked to perform an impromptu analysis when he visited Google. Chief executive Eric Schmidt jokingly asked him for "the most efficient way to sort a million 32-bit integers." Obama had apparently been tipped off, because he quickly replied, "I think the bubble sort would be the wrong way to go." See *http://www.youtube.com/watch?v=k4RRi_ntQc8*.

This is true: bubble sort is conceptually simple but slow for large datasets. The answer Schmidt was probably looking for is "radix sort" (*http://en.wikipedia.org/wiki/Radix_sort*)[1].

The goal of algorithm analysis is to make meaningful comparisons between algorithms, but there are some problems:

---

1. But if you get a question like this in an interview, I think a better answer is, "The fastest way to sort a million integers is to use whatever sort function is provided by the language I'm using. Its performance is good enough for the vast majority of applications, but if it turned out that my application was too slow, I would use a profiler to see where the time was being spent. If it looked like a faster sort algorithm would have a significant effect on performance, then I would look around for a good implementation of radix sort."

- The relative performance of the algorithms might depend on characteristics of the hardware, so one algorithm might be faster on Machine A, another on Machine B. The general solution to this problem is to specify a **machine model** and analyze the number of steps, or operations, an algorithm requires under a given model.

- Relative performance might depend on the details of the dataset. For example, some sorting algorithms run faster if the data are already partially sorted; other algorithms run slower in this case. A common way to avoid this problem is to analyze the **worst case** scenario. It is sometimes useful to analyze average case performance, but that's usually harder, and it might not be obvious what set of cases to average over.

- Relative performance also depends on the size of the problem. A sorting algorithm that is fast for small lists might be slow for long lists. The usual solution to this problem is to express run time (or number of operations) as a function of problem size, and to compare the functions **asymptotically** as the problem size increases.

The good thing about this kind of comparison is that it lends itself to simple classification of algorithms. For example, if I know that the run time of Algorithm A tends to be proportional to the size of the input, n, and Algorithm B tends to be proportional to n^2, then I expect A to be faster than B for large values of n.

This kind of analysis comes with some caveats, but we'll get to that later.

Order of Growth

Suppose you have analyzed two algorithms and expressed their run times in terms of the size of the input: Algorithm A takes *100n+1* steps to solve a problem with size n; Algorithm B takes $n^2 + n + 1$ steps.

The following table shows the run time of these algorithms for different problem sizes:

Input size	Run time of Algorithm A	Run time of Algorithm B
10	1 001	111
100	10 001	10 101
1 000	100 001	1 001 001
10 000	1 000 001	$> 10^{10}$

At *n=10*, Algorithm A looks pretty bad; it takes almost 10 times longer than Algorithm B. But for *n=100* they are about the same, and for larger values A is much better.

The fundamental reason is that for large values of n, any function that contains an n^2 term will grow faster than a function whose leading term is n. The **leading term** is the term with the highest exponent.

For Algorithm A, the leading term has a large coefficient, 100, which is why B does better than A for small n. But regardless of the coefficients, there will always be some value of n where $an^2 > bn$.

The same argument applies to the non-leading terms. Even if the run time of Algorithm A were $n+1000000$, it would still be better than Algorithm B for sufficiently large n.

In general, we expect an algorithm with a smaller leading term to be a better algorithm for large problems, but for smaller problems, there may be a **crossover point** where another algorithm is better. The location of the crossover point depends on the details of the algorithms, the inputs, and the hardware, so it is usually ignored for purposes of algorithmic analysis. But that doesn't mean you can forget about it.

If two algorithms have the same leading order term, it is hard to say which is better; again, the answer depends on the details. So for algorithmic analysis, functions with the same leading term are considered equivalent, even if they have different coefficients.

An **order of growth** is a set of functions whose asymptotic growth behavior is considered equivalent. For example, $2n$, $100n$ and $n+1$ belong to the same order of growth, which is written $O(n)$ in **Big-Oh notation** and often called **linear** because every function in the set grows linearly with n.

All functions with the leading term n^2 belong to $O(n^2)$; they are **quadratic**, which is a fancy word for functions with the leading term n^2.

The following table shows some of the orders of growth that appear most commonly in algorithmic analysis, in increasing order of badness.

Order of growth	Name
$O(1)$	constant
$O(\log_b n)$	logarithmic (for any b)
$O(n)$	linear
$O(n\log_b n)$	"en log en"
$O(n^2)$	quadratic
$O(n^3)$	cubic
$O(c^n)$	exponential (for any c)

For the logarithmic terms, the base of the logarithm doesn't matter; changing bases is the equivalent of multiplying by a constant, which doesn't change the order of growth. Similarly, all exponential functions belong to the same order of growth regardless of the base of the exponent. Exponential functions grow very quickly, so exponential algorithms are only useful for small problems.

Exercise B-1.

Read the Wikipedia page on Big-Oh notation at *http://en.wikipedia.org/wiki/Big_O_notation* and answer the following questions:

1. What is the order of growth of $n^3 + n^2$? What about $1000000n^3 + n^2$? What about $n^3 + 1000000n^2$?

2. What is the order of growth of $(n^2 + n) \cdot (n + 1)$? Before you start multiplying, remember that you only need the leading term.

3. If f is in $O(g)$, for some unspecified function g, what can we say about $af+b$?

4. If f_1 and f_2 are in $O(g)$, what can we say about $f_1 + f_2$?

5. If f_1 is in $O(g)$ and f_2 is in $O(h)$, what can we say about $f_1 + f_2$?

6. If f_1 is in $O(g)$ and f_2 is $O(h)$, what can we say about $f_1 \cdot f_2$?

Programmers who care about performance often find this kind of analysis hard to swallow. They have a point: sometimes the coefficients and the non-leading terms make a real difference. Sometimes the details of the hardware, the programming language, and the characteristics of the input make a big difference. And for small problems asymptotic behavior is irrelevant.

But if you keep those caveats in mind, algorithmic analysis is a useful tool. At least for large problems, the "better" algorithm is usually better, and sometimes it is *much* better. The difference between two algorithms with the same order of growth is usually a constant factor, but the difference between a good algorithm and a bad algorithm is unbounded!

Analysis of Basic Python Operations

Most arithmetic operations are constant time; multiplication usually takes longer than addition and subtraction, and division takes even longer, but these run times don't depend on the magnitude of the operands. Very large integers are an exception; in that case the run time increases with the number of digits.

Indexing operations—reading or writing elements in a sequence or dictionary—are also constant time, regardless of the size of the data structure.

A for loop that traverses a sequence or dictionary is usually linear, as long as all of the operations in the body of the loop are constant time. For example, adding up the elements of a list is linear:

```
total = 0
for x in t:
    total += x
```

The built-in function sum is also linear because it does the same thing, but it tends to be faster because it is a more efficient implementation; in the language of algorithmic analysis, it has a smaller leading coefficient.

If you use the same loop to "add" a list of strings, the run time is quadratic because string concatenation is linear.

The string method join is usually faster because it is linear in the total length of the strings.

As a rule of thumb, if the body of a loop is in $O(n^a)$ then the whole loop is in $O(n^{a+1})$. The exception is if you can show that the loop exits after a constant number of iterations. If a loop runs k times regardless of n, then the loop is in $O(n^a)$, even for large k.

Multiplying by k doesn't change the order of growth, but neither does dividing. So if the body of a loop is in $O(n^a)$ and it runs n/k times, the loop is in $O(n^{a+1})$, even for large k.

Most string and tuple operations are linear, except indexing and len, which are constant time. The built-in functions min and max are linear. The run-time of a slice operation is proportional to the length of the output, but independent of the size of the input.

All string methods are linear, but if the lengths of the strings are bounded by a constant —for example, operations on single characters—they are considered constant time.

Most list methods are linear, but there are some exceptions:

- Adding an element to the end of a list is constant time on average; when it runs out of room it occasionally gets copied to a bigger location, but the total time for n operations is $O(n)$, so we say that the "amortized" time for one operation is $O(1)$.

- Removing an element from the end of a list is constant time.

- Sorting is $O(n\log n)$.

Most dictionary operations and methods are constant time, but there are some exceptions:

- The run time of copy is proportional to the number of elements, but not the size of the elements (it copies references, not the elements themselves).

- The run time of update is proportional to the size of the dictionary passed as a parameter, not the dictionary being updated.

- keys, values and items are linear because they return new lists; iterkeys, iter values and iteritems are constant time because they return iterators. But if you loop through the iterators, the loop will be linear. Using the "iter" functions saves some overhead, but it doesn't change the order of growth unless the number of items you access is bounded.

The performance of dictionaries is one of the minor miracles of computer science. We will see how they work in "Hashtables" (page 249).

Exercise B-2.

Read the Wikipedia page on sorting algorithms at *http://en.wikipedia.org/wiki/ Sorting_algorithm* and answer the following questions:

1. What is a "comparison sort?" What is the best worst-case order of growth for a comparison sort? What is the best worst-case order of growth for any sort algorithm?

2. What is the order of growth of bubble sort, and why does Barack Obama think it is "the wrong way to go?"

3. What is the order of growth of radix sort? What preconditions do we need to use it?

4. What is a stable sort and why might it matter in practice?

5. What is the worst sorting algorithm (that has a name)?

6. What sort algorithm does the C library use? What sort algorithm does Python use? Are these algorithms stable? You might have to Google around to find these answers.

7. Many of the non-comparison sorts are linear, so why does does Python use an $O(n\log n)$ comparison sort?

Analysis of Search Algorithms

A **search** is an algorithm that takes a collection and a target item and determines whether the target is in the collection, often returning the index of the target.

The simplest search algorithm is a "linear search," which traverses the items of the collection in order, stopping if it finds the target. In the worst case it has to traverse the entire collection, so the run time is linear.

The in operator for sequences uses a linear search; so do string methods like find and count.

If the elements of the sequence are in order, you can use a **bisection search**, which is $O(\log n)$. Bisection search is similar to the algorithm you probably use to look a word up in a dictionary (a real dictionary, not the data structure). Instead of starting at the beginning and checking each item in order, you start with the item in the middle and check whether the word you are looking for comes before or after. If it comes before, then you search the first half of the sequence. Otherwise you search the second half. Either way, you cut the number of remaining items in half.

If the sequence has 1,000,000 items, it will take about 20 steps to find the word or conclude that it's not there. So that's about 50,000 times faster than a linear search.

Exercise B-3.
Write a function called bisection that takes a sorted list and a target value and returns the index of the value in the list, if it's there, or None if it's not.

Or you could read the documentation of the bisect module and use that!

Bisection search can be much faster than linear search, but it requires the sequence to be in order, which might require extra work.

There is another data structure, called a **hashtable** that is even faster—it can do a search in constant time—and it doesn't require the items to be sorted. Python dictionaries are implemented using hashtables, which is why most dictionary operations, including the in operator, are constant time.

Hashtables

To explain how hashtables work and why their performance is so good, I start with a simple implementation of a map and gradually improve it until it's a hashtable.

I use Python to demonstrate these implementations, but in real life you wouldn't write code like this in Python; you would just use a dictionary! So for the rest of this chapter, you have to imagine that dictionaries don't exist and you want to implement a data structure that maps from keys to values. The operations you have to implement are:

add(k, v):
    Add a new item that maps from key k to value v. With a Python dictionary, d, this operation is written d[k] = v.

get(target):
    Look up and return the value that corresponds to key target. With a Python dictionary, d, this operation is written d[target] or d.get(target).

For now, I assume that each key only appears once. The simplest implementation of this interface uses a list of tuples, where each tuple is a key-value pair.

```
class LinearMap(object):

    def __init__(self):
        self.items = []

    def add(self, k, v):
        self.items.append((k, v))

    def get(self, k):
        for key, val in self.items:
            if key == k:
                return val
        raise KeyError
```

add appends a key-value tuple to the list of items, which takes constant time.

get uses a for loop to search the list: if it finds the target key it returns the corresponding value; otherwise it raises a KeyError. So get is linear.

An alternative is to keep the list sorted by key. Then get could use a bisection search, which is $O(\log n)$. But inserting a new item in the middle of a list is linear, so this might not be the best option. There are other data structures (see *http://en.wikipedia.org/wiki/Red-black_tree*) that can implement add and get in log time, but that's still not as good as constant time, so let's move on.

One way to improve LinearMap is to break the list of key-value pairs into smaller lists. Here's an implementation called BetterMap, which is a list of 100 LinearMaps. As we'll see in a second, the order of growth for get is still linear, but BetterMap is a step on the path toward hashtables:

```
class BetterMap(object):

    def __init__(self, n=100):
        self.maps = []
        for i in range(n):
            self.maps.append(LinearMap())

    def find_map(self, k):
        index = hash(k) % len(self.maps)
        return self.maps[index]

    def add(self, k, v):
        m = self.find_map(k)
        m.add(k, v)

    def get(self, k):
        m = self.find_map(k)
        return m.get(k)
```

__init__ makes a list of n LinearMaps.

`find_map` is used by `add` and `get` to figure out which map to put the new item in, or which map to search.

`find_map` uses the built-in function `hash`, which takes almost any Python object and returns an integer. A limitation of this implementation is that it only works with hashable keys. Mutable types like lists and dictionaries are unhashable.

Hashable objects that are considered equal return the same hash value, but the converse is not necessarily true: two different objects can return the same hash value.

`find_map` uses the modulus operator to wrap the hash values into the range from 0 to `len(self.maps)`, so the result is a legal index into the list. Of course, this means that many different hash values will wrap onto the same index. But if the hash function spreads things out pretty evenly (which is what hash functions are designed to do), then we expect *n/100* items per LinearMap.

Since the run time of `LinearMap.get` is proportional to the number of items, we expect BetterMap to be about 100 times faster than LinearMap. The order of growth is still linear, but the leading coefficient is smaller. That's nice, but still not as good as a hashtable.

Here (finally) is the crucial idea that makes hashtables fast: if you can keep the maximum length of the LinearMaps bounded, `LinearMap.get` is constant time. All you have to do is keep track of the number of items and when the number of items per LinearMap exceeds a threshold, resize the hashtable by adding more LinearMaps.

Here is an implementation of a hashtable:

```
class HashMap(object):

    def __init__(self):
        self.maps = BetterMap(2)
        self.num = 0

    def get(self, k):
        return self.maps.get(k)

    def add(self, k, v):
        if self.num == len(self.maps.maps):
            self.resize()

        self.maps.add(k, v)
        self.num += 1

    def resize(self):
        new_maps = BetterMap(self.num * 2)

        for m in self.maps.maps:
```

```
        for k, v in m.items:
            new_maps.add(k, v)

    self.maps = new_maps
```

Each HashMap contains a BetterMap; __init__ starts with just 2 LinearMaps and initializes num, which keeps track of the number of items.

get just dispatches to BetterMap. The real work happens in add, which checks the number of items and the size of the BetterMap: if they are equal, the average number of items per LinearMap is 1, so it calls resize.

resize make a new BetterMap, twice as big as the previous one, and then "rehashes" the items from the old map to the new.

Rehashing is necessary because changing the number of LinearMaps changes the denominator of the modulus operator in find_map. That means that some objects that used to wrap into the same LinearMap will get split up (which is what we wanted, right?).

Rehashing is linear, so resize is linear, which might seem bad, since I promised that add would be constant time. But remember that we don't have to resize every time, so add is usually constant time and only occasionally linear. The total amount of work to run add n times is proportional to n, so the average time of each add is constant time!

To see how this works, think about starting with an empty HashTable and adding a sequence of items. We start with 2 LinearMaps, so the first 2 adds are fast (no resizing required). Let's say that they take one unit of work each. The next add requires a resize, so we have to rehash the first two items (let's call that 2 more units of work) and then add the third item (one more unit). Adding the next item costs 1 unit, so the total so far is 6 units of work for 4 items.

The next add costs 5 units, but the next three are only one unit each, so the total is 14 units for the first 8 adds.

The next add costs 9 units, but then we can add 7 more before the next resize, so the total is 30 units for the first 16 adds.

After 32 adds, the total cost is 62 units, and I hope you are starting to see a pattern. After n adds, where n is a power of two, the total cost is $2n-2$ units, so the average work per add is a little less than 2 units. When n is a power of two, that's the best case; for other values of n the average work is a little higher, but that's not important. The important thing is that it is $O(1)$.

Figure B-1 shows how this works graphically. Each block represents a unit of work. The columns show the total work for each add in order from left to right: the first two adds cost 1 units, the third costs 3 units, etc.

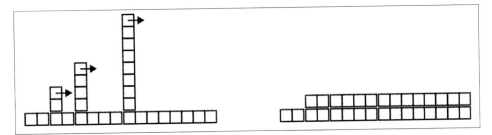

Figure B-1. The cost of a hashtable add.

The extra work of rehashing appears as a sequence of increasingly tall towers with increasing space between them. Now if you knock over the towers, amortizing the cost of resizing over all adds, you can see graphically that the total cost after n adds is $2n - 2$.

An important feature of this algorithm is that when we resize the HashTable it grows geometrically; that is, we multiply the size by a constant. If you increase the size arithmetically—adding a fixed number each time—the average time per add is linear.

You can download my implementation of HashMap from *http://thinkpython.com/code/Map.py*, but remember that there is no reason to use it; if you want a map, just use a Python dictionary.

APPENDIX C
Lumpy

Throughout the book, I have used diagrams to represent the state of running programs.

In "Variables" (page 14), we used a state diagram to show the names and values of variables. In "Stack Diagrams" (page 30) I introduced a stack diagram, which shows one frame for each function call. Each frame shows the parameters and local variables for the function or method. Stack diagrams for recursive functions appear in "Stack Diagrams for Recursive Functions" (page 54) and "More Recursion" (page 66).

"Lists Are Mutable" (page 106) shows what a list looks like in a state diagram, "Dictionaries and Lists" (page 126) shows what a dictionary looks like, and "Dictionaries and Tuples" (page 139) shows two ways to represent tuples.

"Attributes" (page 172) introduces object diagrams, which show the state of an object's attributes, and their attributes, and so on. "Rectangles" (page 173) has object diagrams for Rectangles and their embedded Points. "Time" (page 181) shows the state of a Time object. "Class Attributes" (page 202) has a diagram that includes a class object and an instance, each with their own attributes.

Finally, "Class Diagrams" (page 209) introduces class diagrams, which show the classes that make up a program and the relationships between them.

These diagrams are based on the Unified Modeling Language (UML), which is a standardized graphical language used by software engineers to communicate about program design, especially for object-oriented programs.

UML is a rich language with many kinds of diagrams that represent many kinds of relationship between objects and classes. What I presented in this book is a small subset of the language, but it is the subset most commonly used in practice.

The purpose of this appendix is to review the diagrams presented in the previous chapters, and to introduce Lumpy. Lumpy, which stands for "UML in Python," with some of the letters rearranged, is part of Swampy, which you already installed if you worked on the case study in Chapter 4 or Chapter 19, or if you did Exercise 15-4,

Lumpy uses Python's inspect module to examine the state of a running program and generate object diagrams (including stack diagrams) and class diagrams.

State Diagram

Here's an example that uses Lumpy to generate a state diagram.

```
from swampy.Lumpy import Lumpy

lumpy = Lumpy()
lumpy.make_reference()

message = 'And now for something completely different'
n = 17
pi = 3.1415926535897932

lumpy.object_diagram()
```

The first line imports the Lumpy class from swampy.Lumpy. If you don't have Swampy installed as a package, make sure the Swampy files are in Python's search path and use this import statement instead:

```
from Lumpy import Lumpy
```

The next lines create a Lumpy object and make a "reference" point, which means that Lumpy records the objects that have been defined so far.

Next we define new variables and invoke object_diagram, which draws the objects that have been defined since the reference point, in this case message, n and pi.

Figure C-1 shows the result. The graphical style is different from what I showed earlier; for example, each reference is represented by a circle next to the variable name and a line to the value. And long strings are truncated. But the information conveyed by the diagram is the same.

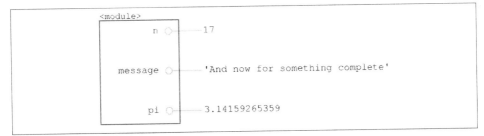

Figure C-1. State diagram generated by Lumpy.

The variable names are in a frame labeled <module>, which indicates that these are module-level variables, also known as global.

You can download this example from *http://thinkpython.com/code/lumpy_demo1.py*. Try adding some additional assignments and see what the diagram looks like.

Stack Diagram

Here's an example that uses Lumpy to generate a stack diagram. You can download it from here (*http://thinkpython.com/code/lumpy_demo2.py*).

```
from swampy.Lumpy import Lumpy

def countdown(n):
    if n <= 0:
        print 'Blastoff!'
        lumpy.object_diagram()
    else:
        print n
        countdown(n-1)

lumpy = Lumpy()
lumpy.make_reference()
countdown(3)
```

Figure C-2 shows the result. Each frame is represented with a box that has the function's name outside and variables inside. Since this function is recursive, there is one frame for each level of recursion.

Figure C-2. Stack diagram.

Remember that a stack diagram shows the state of the program at a particular point in its execution. To get the diagram you want, sometimes you have to think about where to invoke object_diagram.

In this case I invoke `object_diagram` after executing the base case of the recursion; that way the stack diagram shows each level of the recursion. You can call `object_dia gram` more than once to get a series of snapshots of the program's execution.

Object Diagrams

This example generates an object diagram showing the lists from "A List Is a Sequence" (page 105). You can download it from *http://thinkpython.com/code/lumpy_demo3.py*.

```
from swampy.Lumpy import Lumpy

lumpy = Lumpy()
lumpy.make_reference()

cheeses = ['Cheddar', 'Edam', 'Gouda']
numbers = [17, 123]
empty = []

lumpy.object_diagram()
```

Figure C-3 shows the result. Lists are represented by a box that shows the indices mapping to the elements. This representation is slightly misleading, since indices are not actually part of the list, but I think they make the diagram easier to read. The empty list is represented by an empty box.

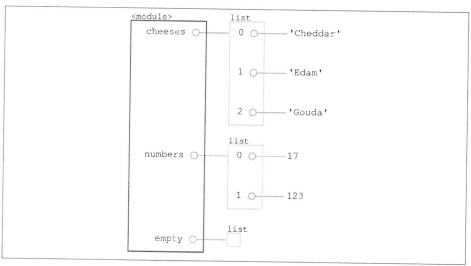

Figure C-3. Object diagram.

And here's an example showing the dictionaries from "Dictionaries and Lists" (page 126). You can download it from *http://thinkpython.com/code/lumpy_demo4.py*.

```
from swampy.Lumpy import Lumpy

lumpy = Lumpy()
lumpy.make_reference()

hist = histogram('parrot')
inverse = invert_dict(hist)

lumpy.object_diagram()
```

Figure C-4 shows the result. hist is a dictionary that maps from characters (single-letter strings) to integers; inverse maps from integers to lists of strings.

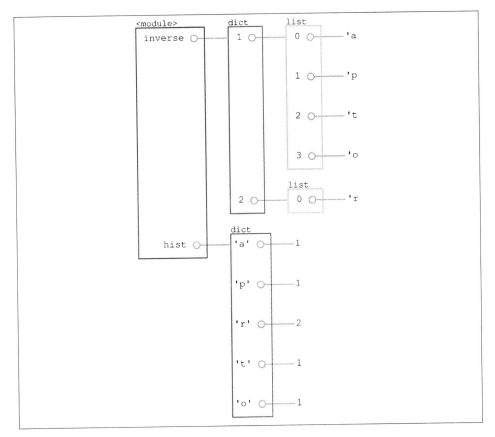

Figure C-4. Object diagram.

This example generates an object diagram for Point and Rectangle objects, as in "Copying" (page 176). You can download it from *http://thinkpython.com/code/lumpy_demo5.py.*

```
import copy
from swampy.Lumpy import Lumpy

lumpy = Lumpy()
lumpy.make_reference()

box = Rectangle()
box.width = 100.0
box.height = 200.0
box.corner = Point()
box.corner.x = 0.0
box.corner.y = 0.0

box2 = copy.copy(box)

lumpy.object_diagram()
```

Figure C-5 shows the result. copy.copy make a shallow copy, so box and box2 have their own width and height, but they share the same embedded Point object. This kind of sharing is usually fine with immutable objects, but with mutable types, it is highly error-prone.

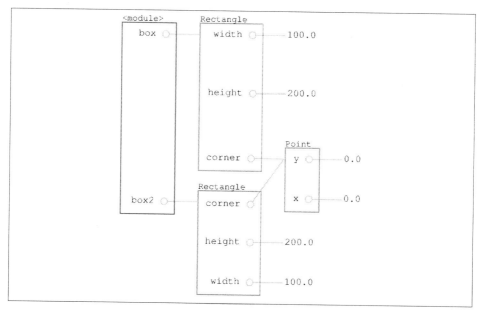

Figure C-5. Object diagram.

Function and Class Objects

When I use Lumpy to make object diagrams, I usually define the functions and classes before I make the reference point. That way, function and class objects don't appear in the diagram.

But if you are passing functions and classes as parameters, you might want them to appear. This example shows what that looks like; you can download it from *http://thinkpython.com/code/lumpy_demo6.py*.

```python
import copy
from swampy.Lumpy import Lumpy

lumpy = Lumpy()
lumpy.make_reference()

class Point(object):
    """Represents a point in 2-D space."""

class Rectangle(object):
    """Represents a rectangle."""

def instantiate(constructor):
    """Instantiates a new object."""
    obj = constructor()
    lumpy.object_diagram()
    return obj

point = instantiate(Point)
```

Figure C-6 shows the result. Since we invoke `object_diagram` inside a function, we get a stack diagram with a frame for the module-level variables and for the invocation of `instantiate`.

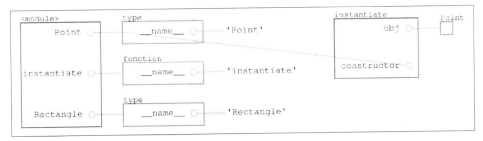

Figure C-6. Object diagram.

At the module level, `Point` and `Rectangle` refer to class objects (which have type `type`); `instantiate` refers to a function object.

This diagram might clarify two points of common confusion: (1) the difference between the class object, `Point`, and the instance of Point, `obj`, and (2) the difference between the function object created when `instantiate` is defined, and the frame created with it is called.

Class Diagrams

Although I distinguish between state diagrams, stack diagrams and object diagrams, they are mostly the same thing: they show the state of a running program at a point in time.

Class diagrams are different. They show the classes that make up a program and the relationships between them. They are timeless in the sense that they describe the program as a whole, not any particular point in time. For example, if an instance of Class A generally contains a reference to an instance of Class B, we say there is a "HAS-A relationship" between those classes.

Here's an example that shows a HAS-A relationship. You can download it from *http://thinkpython.com/code/lumpy_demo7.py*.

```
from swampy.Lumpy import Lumpy

lumpy = Lumpy()
lumpy.make_reference()

box = Rectangle()
box.width = 100.0
box.height = 200.0
box.corner = Point()
box.corner.x = 0.0
box.corner.y = 0.0

lumpy.class_diagram()
```

Figure C-7 shows the result. Each class is represented with a box that contains the name of the class, any methods the class provides, any class variables, and any instance variables. In this example, `Rectangle` and `Point` have instance variables, but no methods or class variables.

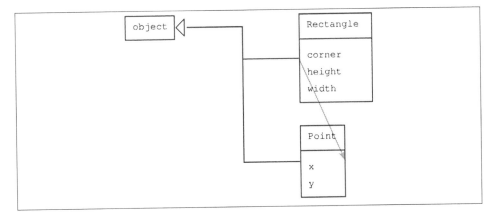

Figure C-7. Class diagram.

The arrow from `Rectangle` to `Point` shows that Rectangles contain an embedded Point. In addition, `Rectangle` and `Point` both inherit from `object`, which is represented in the diagram with a triangle-headed arrow.

Here's a more complex example using my solution to Exercise 18-6. You can download the code from *http://thinkpython.com/code/lumpy_demo8.py*; you will also need *http://thinkpython.com/code/PokerHand.py.*

```
from swampy.Lumpy import Lumpy

from PokerHand import *

lumpy = Lumpy()
lumpy.make_reference()

deck = Deck()
hand = PokerHand()
deck.move_cards(hand, 7)

lumpy.class_diagram()
```

Figure C-8 shows the result. `PokerHand` inherits from `Hand`, which inherits from `Deck`. Both `Deck` and `PokerHand` have Cards.

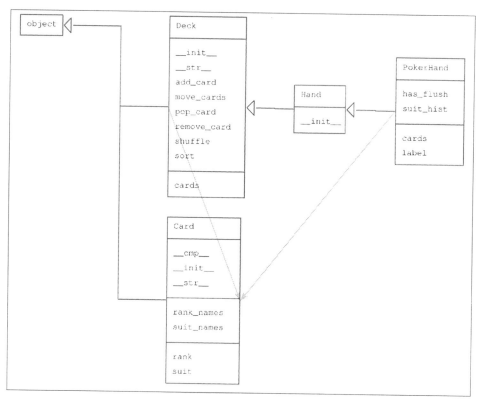

Figure C-8. Class diagram.

This diagram does not show that Hand also has cards, because in the program there are no instances of Hand. This example demonstrates a limitation of Lumpy; it only knows about the attributes and HAS-A relationships of objects that are instantiated.

Index

A

abecedarian, 87, 99
abs function, 62
absolute path, 161, 168
access, 106
accumulator, 117
    histogram, 150
    list, 110
    string, 206
    sum, 109
Ackermann function, 72, 129
add method, 194
addition with carrying, 80
algorithm, 4, 10, 80, 153, 243
    Euclid, 73
    MD5, 166
    RSA, 131
    square root, 82
aliasing, 112, 113, 118, 173, 176, 199
    copying to avoid, 116
alphabet, 47
alternative execution, 51
ambiguity, 6
anagram, 118
anagram set, 145, 165
analysis of algorithms, 243
analysis of primitives, 246
and operator, 50

anydbm module, 163
append method, 108, 115, 119, 205, 206
arc function, 40
Archimedian spiral, 47
argument, 23, 26, 28, 29, 34, 114
    gather, 137
    keyword, 41, 45, 142, 218
    list, 114
    optional, 90, 112, 125
    variable-length tuple, 137
argument scatter, 138
arithmetic operator, 16
assert statement, 186
assignment, 21, 75, 105
    augmented, 109, 117
    item, 88, 106, 136
    multiple, 81, 129
    tuple, 136, 137, 138, 144
assignment statement, 14
asymptotic analysis, 244
attribute, 197
    class, 202, 212
    initializing, 197
    instance, 172, 178, 203, 212
    __dict__, 197
AttributeError, 177, 238
augmented assignment, 109, 117
Austin, Jane, 149
available colors, 179, 200

average case, 244
average cost, 252

B

badness, 245
Bangladesh, national flag, 179
base case, 54, 58, 258
benchmarking, 155, 157
BetterMap, 250
big, hairy expression, 240
big-oh notation, 245
binary search, 119
binding, 225, 229
bingo, 145
birthday, 187
birthday paradox, 118
bisect module, 249
bisect module, 119, 249
bisection search, 119, 248
bisection, debugging by, 81
bitwise operator, 16
body, 26, 33, 77
bool type, 50
boolean expression, 49, 57
boolean function, 65, 181
boolean operator, 91
borrowing, subtraction with, 80, 185
bound method, 223, 229
bounded, 251
bounding box, 179, 220, 229
bracket
    squiggly, 121
bracket operator, 85, 106, 136
branch, 51, 58
break statement, 78
bubble sort, 243
bug, 4, 4, 10
    worst, 199
    worst ever, 231
Button widget, 218

C

calculator, 11, 22
call graph, 128, 133
Callable object, 225
callback, 219, 223, 225, 225, 227, 229
Canvas coordinate, 220, 227
Canvas item, 220

Canvas object, 179
Canvas widget, 219
Car Talk, 103, 103, 103, 133, 145
Card class, 202, 263
card, playing, 201
carrying, addition with, 80, 183, 184
case-sensitivity, variable names, 20
catch, 169
chained conditional, 51, 58
character, 85
checksum, 166, 166
child class, 207, 212
choice function, 149
circle function, 40
circular definition, 66
class, 171, 178
    Card, 202
    child, 207, 212
    Deck, 205
    Hand, 207
    Kangaroo, 199
    parent, 207
    Point, 171, 193, 259
    Rectangle, 174, 259
    SimpleTurtleWorld, 222
    Time, 181
class attribute, 202, 212
class definition, 171
class diagram, 209, 213, 255, 262, 262
class object, 172, 178, 261
close method, 160, 164, 165
cmp function, 205
__cmp__ method, 204
Collatz conjecture, 78
colon, 26, 234
color list, 179, 200
comment, 19, 21
commutativity, 19, 195
compare function, 62
comparing algorithms, 243
comparison
    string, 92
    tuple, 141, 205
comparison sort, 248
compile, 2, 9
composition, 25, 28, 34, 64, 205
compound statement, 51, 58
concatenation, 18, 21, 29, 87, 89, 112
    list, 107, 115, 119

global variable, 129, 133, 257
    update, 130
GNU Free Documentation License, 12, 13
graphical user interface, 217
greatest common divisor (GCD), 73
grid, 36
guardian pattern, 70, 72, 92
GUI, 217, 229
Gui module, 217

H

Hand class, 207, 263
hanging, 235
HAS-A relationship, 209, 213, 262
hasattr function, 177, 197
hash function, 127, 132, 251
hashable, 127, 132, 140
HashMap, 251
hashtable, 122, 132, 249
header, 26, 33, 234
Hello, World, 7
help utility, 11
hexadecimal, 172
high-level language, 1, 9
histogram, 123, 123, 133
    random choice, 149, 152
    word frequencies, 149
Holmes, Sherlock, 5
homophone, 134
HTMLParser module, 231
hyperlink, 231
hypotenuse, 64

I

identical, 118
identity, 113
if statement, 50
Image module, 230
image viewer, 230
immutability, 88, 88, 94, 114, 127, 135, 142
implementation, 123, 132, 155, 197
import statement, 34, 38, 167
in operator, 248
in operator, 91, 107, 122
increment, 76, 82, 183, 191
incremental development, 71, 234
indentation, 26, 190, 234

index, 85, 85, 92, 94, 106, 117, 121, 237
    looping with, 100, 107
    negative, 86
    slice, 87, 108
    starting at zero, 85, 106
IndexError, 86, 93, 107, 238
indexing, 246
infinite loop, 77, 82, 217, 235, 236
infinite recursion, 55, 58, 69, 235, 236
information hiding, 198, 199
inheritance, 207, 212
init method, 192, 197, 202, 205, 207
initialization
    variable, 81
initialization (before update), 76
instance, 38, 45, 172, 178
    as argument, 173
    as return value, 174
instance attribute, 172, 178, 203, 212
instantiate, 261
instantiation, 172
int function, 23
int type, 13
integer, 20
    long, 130
interactive mode, 2, 9, 17, 31
interface, 42, 45, 45, 197, 210
interlocking words, 119
interpret, 2, 9
invariant, 185, 187, 228
invert dictionary, 126
invocation, 90, 95
IOError, 162
is operator, 113, 176
IS-A relationship, 209, 212, 263
isinstance function, 69, 195
item, 94, 105
    Canvas, 220, 229
    dictionary, 132
item assignment, 88, 106, 136
item update, 107
items method, 139
iteration, 76, 82

J

join, 247
join method, 112, 206

membership
    binary search, 119
    bisection search, 119
    dictionary, 122
    list, 107
    set, 122
memo, 128, 133
mental model, 239
Menubutton widget, 224
metaphor, method invocation, 191
metathesis, 145
method, 90, 95, 189, 198
    add, 194
    append, 108, 115, 205, 206
    close, 160, 164, 165
    config, 219
    count, 91
    extend, 109
    get, 124
    init, 192, 202, 205, 207
    items, 139
    join, 112, 206
    keys, 124
    mro, 210
    pop, 111, 206
    radd, 195
    read, 165
    readline, 97, 165
    remove, 111
    replace, 147
    setdefault, 127
    sort, 109, 116, 141, 207
    split, 112, 136
    string, 91
    strip, 98, 147
    translate, 147
    update, 140
    values, 122
    void, 109
    __cmp__, 204
    __str__, 193, 205
method append, 119
method resolution order, 210
method syntax, 191
method, bound, 223
method, list, 108
Meyers, Chris, 13
min function, 137, 138
Moby Project, 97

model, mental, 239
modifier, 183, 186
module, 24, 34, 34
    anydbm, 163
    bisect, 119, 249
    copy, 176
    datetime, 187
    Gui, 217
    HTMLParser, 231
    Image, 230
    os, 161
    pickle, 159, 164
    pprint, 132
    profile, 155
    random, 118, 142, 148, 206
    reload, 167, 235
    shelve, 165
    string, 147
    structshape, 143
    time, 119
    urllib, 169, 231
    Visual, 199
    vpython, 199
    World, 178
module object, 24, 167
module, writing, 166
module-level variable, 257
modulus operator, 49, 57
Monty Python and the Holy Grail, 182
MP3, 166
mro method, 210
multiline string, 44, 234
multiple assignment, 75, 81, 129
multiplicity (in class diagram), 209, 213
mutability, 88, 106, 108, 114, 130, 135, 142, 175
mutable object, as default value, 199

N

NameError, 29, 237
natural language, 6, 10
negative index, 86
nested conditional, 52, 58
nested list, 105, 107, 117
newline, 56, 75, 206
Newton's method, 79
None special value, 31, 62, 71, 109, 111
not operator, 50
number, random, 148

O

Obama, Barack, 243
object, 88, 94, 112, 113, 117, 171
    Callable, 225
    Canvas, 179
    class, 172, 178, 261
    copying, 176
    embedded, 174, 178, 199
    Event, 226
    file, 97, 102
    function, 26, 35, 261
    module, 167
    mutable, 175
    printing, 190
object code, 2, 9
object diagram, 172, 174, 176, 178, 181, 203, 255, 258
object-oriented design, 197
object-oriented language, 198
object-oriented programming, 189, 198, 207
octal, 15
odometer, 103
Olin College, 12
open function, 97, 98, 159, 162, 163
operand, 16, 21
operator, 21
    and, 50
    bitwise, 16
    boolean, 91
    bracket, 85, 106, 136
    del, 111
    format, 160, 168, 237
    in, 91, 107, 122
    is, 113, 176
    logical, 49, 50
    modulus, 49, 57
    not, 50
    or, 50
    overloading, 199
    relational, 50, 204
    slice, 87, 95, 108, 115, 136
    string, 18
    update, 109
operator overloading, 194, 204
operator, arithmetic, 16
option, 218, 229
optional argument, 90, 112, 125
optional parameter, 151, 193
or operator, 50

order of growth, 244
order of operations, 18, 20, 240
os module, 161
other (parameter name), 192
OverflowError, 57
overloading, 199
override, 151, 157, 193, 204, 207, 210

P

package, 37
packing widgets, 223, 229
palindrome, 73, 95, 101, 103, 103
parameter, 28, 29, 33, 114
    gather, 137
    optional, 151, 193
    other, 192
    self, 191
parent class, 207, 207, 212
parentheses
    argument in, 23
    empty, 26, 90
    matching, 4
    overriding precedence, 18
    parameters in, 28, 29
    parent class in, 207
    tuples in, 135
parse, 6, 10
pass statement, 51
path, 161, 168
    absolute, 161
    relative, 161
pattern
    decorate-sort-undecorate, 141
    DSU, 141, 150
    filter, 110, 117
    guardian, 70, 72, 92
    map, 110, 117
    reduce, 109, 117
    search, 89, 94, 99, 125
    swap, 136
pdb (Python debugger), 238
PEMDAS, 18
permission, file, 162
persistence, 159, 168
pi, 25, 83
pickle module, 159, 164
pickling, 164
pie, 47
PIL (Python Imaging Library), 230

pipe, 165
pixel coordinate, 227
plain text, 97, 147, 231
planned development, 184, 186
playing card, Anglo-American, 201
poetry, 7
Point class, 171, 193, 259
point, mathematical, 171
poker, 201, 213
polygon function, 40
polymorphism, 196, 199, 210
pop method, 111, 206
popen function, 165
portability, 1, 9
postcondition, 45, 70, 210
pprint module, 132
practical analysis of algorithms, 246
precedence, 21, 240
precondition, 45, 46, 46, 70, 118, 210
prefix, 153
pretty print, 132
print function, 8
print statement, 8, 10, 193, 238
problem recognition, 100, 101, 102
problem solving, 1, 9
profile module, 155
program, 3, 10
program testing, 102
programming language, 1
Project Gutenberg, 147
prompt, 3, 9, 56
prose, 7
prototype and patch, 182, 184, 186
pseudorandom, 148, 157
pure function, 182, 186
Puzzler, 103, 103, 103, 133, 145
Pythagorean theorem, 62
Python 3, 8, 16, 55, 131, 138
Python debugger (pdb), 238
Python Imaging Library (PIL), 230

Q

quadratic growth, 245
quotation mark, 8, 13, 13, 44, 88, 234

R

radd method, 195
radian, 24

radix sort, 243
rage, 241
raise statement, 125, 186
Ramanujan, Srinivasa, 83
randint function, 118, 148
random function, 142, 148
random module, 118, 142, 148, 206
random number, 148
random text, 153
random walk programming, 157, 241
rank, 201
raw_input function, 55
read method, 165
readline method, 97, 165
Rectangle class, 174, 259
recursion, 53, 53, 58, 66, 68, 257
    base case, 54
    infinite, 55, 69, 236
recursive definition, 66, 146
red-black tree, 250
reduce pattern, 109, 117
reducible word, 134, 146
redundancy, 6
refactoring, 43, 43, 46, 211
reference, 114, 114, 118
    aliasing, 113
rehashing, 252
relational operator, 50, 204
relative path, 161, 168
reload function, 167, 235
remove method, 111
repetition, 38
    list, 108
replace method, 147
repr function, 168
representation, 171, 173, 201
return statement, 54, 61, 240
return value, 23, 34, 61, 174
    tuple, 137
reverse lookup, dictionary, 125, 133
reverse word pair, 119
reversed function, 143
rotation
    letters, 133
rotation, letter, 96
RSA algorithm, 131
rules of precedence, 18, 21
running pace, 11, 22, 187
runtime error, 4, 20, 55, 57, 233, 237

RuntimeError, 55, 69

S

safe language, 4
sanity check, 132
scaffolding, 64, 72, 132
scatter, 138, 144
Schmidt, Eric, 243
Scrabble, 145
script, 3, 9
script mode, 2, 10, 17, 31
search, 125, 248
search pattern, 89, 94, 99
search, binary, 119
search, bisection, 119
secret exercise, 169
self (parameter name), 191
semantic error, 5, 10, 14, 20, 93, 233, 239
semantics, 5, 10, 190
sequence, 85, 94, 105, 112, 135, 142
    coordinate, 220
set, 152
    anagram, 145, 165
set membership, 122
setdefault method, 127
sexagesimal, 184
shallow copy, 176, 178, 260
shape, 144
shape error, 143
shell, 165
shelve module, 165
shuffle function, 206
SimpleTurtleWorld class, 222
sine function, 24
singleton, 126, 133, 135
slice, 94
    copy, 88, 108
    list, 108
    string, 87
    tuple, 136
    update, 108
slice operator, 87, 95, 108, 115, 136
sort method, 109, 116, 141, 207
sorted function, 143
sorting, 247, 248
source code, 2, 9
special case, 102, 102, 183
special value
    False, 50

None, 31, 62, 71, 109, 111
    True, 50
spiral, 47
split method, 112, 136
sqrt, 63
sqrt function, 25
square root, 79
squiggly bracket, 121
stable sort, 248
stack diagram, 30, 30, 34, 46, 54, 67, 72, 115, 255, 257
state diagram, 14, 21, 75, 93, 106, 113, 114, 126, 140, 172, 174, 176, 181, 203, 255, 256
statement, 20
    assert, 186
    assignment, 14, 75
    break, 78
    compound, 51
    conditional, 50, 58, 65
    for, 39, 86, 107
    global, 130
    if, 50
    import, 34, 38, 167
    pass, 51
    print, 8, 10, 193, 238
    raise, 125, 186
    return, 54, 61, 240
    try, 162
    while, 76
step size, 95
str function, 24
__str__ method, 193, 205
string, 13, 20, 112, 142
    accumulator, 206
    comparison, 92
    empty, 112
    immutable, 88
    method, 90
    multiline, 44, 234
    operation, 18
    slice, 87
    triple-quoted, 44
string concatenation, 247
string method, 91
string methods, 247
string module, 147
string representation, 168, 193
string type, 13
strip method, 98, 147

About the Author

Allen Downey is a Professor of Computer Science at Olin College of Engineering. He has taught computer science at Wellesley College, Colby College, and U.C. Berkeley. He has a Ph.D. in Computer Science from U.C. Berkeley and Master's and Bachelor's degrees from MIT.

Colophon

The animal on the cover of *Think Python* is the Carolina parrot, also known as the Carolina parakeet (*Conuropsis carolinensis*). This parrot inhabited the southeastern United States and was the only continental parrot with a habitat north of Mexico. At one time, it lived as far north as New York and the Great Lakes, though it was chiefly found from Florida to the Carolinas.

The Carolina parrot was mainly green with a yellow head and some orange coloring that appeared on the forehead and cheeks at maturity. Its average size ranged from 31–33 cm. It had a loud, riotous call and would chatter constantly while feeding. It inhabited tree hollows near swamps and riverbanks. The Carolina parrot was a very gregarious animal, living in small groups that could grow to several hundred parrots when feeding.

These feeding areas were, unfortunately, often the crops of farmers, who would shoot the birds to keep them away from the harvest. The birds' social nature caused them to fly to the rescue of any wounded parrot, allowing farmers to shoot down whole flocks at a time. In addition, their feathers were used to embellish ladies' hats, and some parrots were kept as pets. A combination of these factors led the Carolina parrot to become rare by the late 1800s, and poultry disease may have contributed to their dwindling numbers. By the 1920s, the species was extinct.

Today, there are more than 700 Carolina parrot specimens preserved in museums worldwide.

The cover image is from *Johnson's Natural History*. The cover font is Adobe ITC Garamond. The text font is Minion Pro; the heading font is Myriad Pro; and the code font is Ubuntu Mono.

Have it your way.

Get even more for your money.

Join the O'Reilly Community, and register the O'Reilly books you own. It's free, and you'll get:

- $4.99 ebook upgrade offer
- 40% upgrade offer on O'Reilly print books
- Membership discounts on books and events
- Free lifetime updates to ebooks and videos
- Multiple ebook formats, DRM FREE
- Participation in the O'Reilly community
- Newsletters
- Account management
- 100% Satisfaction Guarantee

Signing up is easy:

1. **Go to: oreilly.com/go/register**
2. **Create an O'Reilly login.**
3. **Provide your address.**
4. **Register your books.**

Note: English-language books only

To order books online:
oreilly.com/store

For questions about products or an order:
orders@oreilly.com

To sign up to get topic-specific email announcements and/or news about upcoming books, conferences, special offers, and new technologies:
elists@oreilly.com

For technical questions about book content:
booktech@oreilly.com

To submit new book proposals to our editors:
proposals@oreilly.com

O'Reilly books are available in multiple DRM-free ebook formats. For more information:
oreilly.com/ebooks

O'REILLY®

Spreading the knowledge of innovators | oreilly.com

CPSIA information can be obtained at www.ICGtesting.com
Printed in the USA
BVOW062147300912

301647BV00002B/3/P